T0140293

Studies in Computational Intelligence

Volume 486

Series Editor

J. Kacprzyk, Warsaw, Poland

For further volumes:
http://www.springer.com/series/7092

Studies in Computational Intelligence

Volume 486

Barna Iantovics · Roumen Kountchev
Editors

Advanced Intelligent Computational Technologies and Decision Support Systems

 Springer

Editors
Barna Iantovics
Petru Maior University
Targu Mures
Romania

Roumen Kountchev
Technical University of Sofia
Sofia
Bulgaria

ISSN 1860-949X
ISBN 978-3-319-03387-7
DOI 10.1007/978-3-319-00467-9
Springer Cham Heidelberg New York Dordrecht London

ISSN 1860-9503 (electronic)
ISBN 978-3-319-00467-9 (eBook)

Printed on acid-free paper

Springer is part of Springer Science+Business Media (www.springer.com)

Preface

The topic of this book is important. Many real-life problems/tasks are difficult to be solved by humans due to their: complexity/difficulty, large number, and/or the required physical human effort. Many situations in which such problems appear are characteristic to field like healthcare that represented the principal target of our book. Difficult problems solving by humans may involve vagueness, imprecision, and/or uncertainty. Among the examples of difficult problem, we mention: the diagnosis of a very difficult medical case; prescription of a medicine for treating a patient who suffers from a very little known illness whose evolution is unpredictable; analyzing a large number of medical images, signals or multimedia etc. To exemplify situations when an increased human effort is necessary, we mention: search for patients' data in a large number of paper-based medical records, a longer time checking/monitoring of a patient health status, etc.

The previously mentioned motivations suggest the necessity of human specialists to require assistance from computing systems. "Soft" assistance may vary from simple information support (data obtained by request from a database for example) to complex forms of decision support (offered by new generations of intelligent expert systems for example). "Hard" assistance could be offered by systems that include software (for decisions and/or control) and hardware components (for example, a wheelchair of a neurologically disabled patient that must be controlled by software).

The main aim of this book was to offer a state-of-the-art chapters' collection that cover themes related to *Advanced Intelligent Computational Technologies and Decision Support Systems* that could be applied for different problems solving in fields like healthcare assisting the humans in solving problems. Efficient solving of various problems by computing systems may require different problem solving algorithms adapted based on the specific of the problems type, complexity and the information/details known about the problems. Computational intelligence algorithms are frequently used for different computational hard problems solving. Computational intelligence provides algorithms for important applications in medicine, such as diagnosis, health datasets analyzing, and drug design. Sometime the combination/hybridization of more algorithms is necessary, in order to obtain their combined advantages. The combination of a neural network with a genetic algorithm represents such an example of hybridization.

The book brings forward a wealth of ideas, algorithms, and case studies in themes like: intelligent predictive diagnosis; intelligent analyzing of medical images; new format for coding of single and sequences of medical images; Medical Decision Support Systems; diagnosis of Down's syndrome; computational perspectives for electronic fetal monitoring; efficient compression of CT Images; adaptive interpolation and halftoning for medical images; development of a brain–computer interface decision support system for controlling the movement of a wheelchair for neurologically disabled patients using their electroencephalography; applications of artificial neural networks for real-life problems solving; present and perspectives for Electronic Healthcare Record Systems; adaptive approaches for noise reduction in sequences of CT images; rule-based classification of patients screened with the MMPI test; scan converting OCT images using Fourier analysis; teaching for long-term memory; Medical Informatics; Intelligent Mobile Multiagent Systems; quantifying anticipatory characteristics, anticipation scope and anticipatory profile; Bio-inspired Computational Methods; Complex Systems; optimization problems solving; fast cost update and query algorithms; new generation of biomedical equipment based on FPGA; negotiation-based patient scheduling in hospitals, etc.

July 2013

Barna Iantovics
Roumen Kountchev

Acknowledgments

The chapters included in this book were selected form contributions to the first Edition of *International Workshop on Next Generation Intelligent Medical Decision Support Systems*, *MedDecSup* 2011 held on September 2011 at Petru Maior University, organized by Petru Maior University, Tirgu-Mures, Romania and the Technical University of Sofia, Bulgaria. *MedDecSup* 2011 was organized under the frame of bilateral Romanian–Bulgarian cooperation research project with a duration of two years that bears the title "Electronic Health Records for the Next Generation Medical Decision Support in Romanian and Bulgarian National Healthcare Systems", Acronym: NextGenElectroMedSupport, National Authority for Scientific Research (ANCS), Romania and Executive Unit for Financing Higher Education, Research, Development and Innovation (UEFISCU) and the Ministry of Education, Youth and Science of Bulgaria, having as directors from Bulgaria Prof. Roumen Kountchev, Technical University of Sofia and from Romania Dr. Barna Iantovics, Petru Maior University.

The research of Barna Iantovics was achieved within the project "Transnational Network for Integrated Management of Postdoctoral Research in Communicating Sciences. Institutional building (postdoctoral school) and fellowships program (CommScie)"—POSDRU/89/1.5/S/63663, Sectorial Operational Programme Human Resources Development 2007−2013.

The chapters of the workshop were useful for the bilateral cooperation research project between Romania and Slovakia with the duration of two years that bears the title: "Hybrid Medical Complex Systems", Acronym: ComplexMediSys (2011−2012).

Contents

Medical Decision Support System Using Pattern Recognition Methods for Assessment of Dermatoglyphic Indices and Diagnosis of Down's Syndrome

Hubert Wojtowicz and Wieslaw Wajs

Abstract The development and implementation of the telemedical system for the diagnosis of Down's syndrome is described in the chapter. The system is a tool supporting medical decision by automatic processing of dermatoglyphic prints and detecting features indicating the presence of genetic disorder. The application of image processing methods for the pre-processing and enhancement of dermato-glyphic images has also been presented. Classifiers for the recognition of finger-print patterns and patterns of the hallucal area of the soles, which are parts of an automatic system for rapid screen diagnosing of trisomy 21 (Down's Syndrome) in infants, are created and discussed. The method and algorithms for the calculation of palmprint's ATD angle are presented then. The images of dermatoglyphic prints are pre-processed before the classification stage to extract features analyzed by Support Vector Machines algorithm. The application of an algorithm based on multi-scale pyramid decomposition of an image is proposed for ridge orientation calculation. RBF and triangular kernel types are used in training of SVM multi-class systems generated with one-vs.-one scheme. A two stage algorithm for the calculation of palmprint's singular points location, based on improved Poincare index and Gaussian-Hermite moments is subsequently discussed. The results of experiments conducted on the database of Collegium Medicum of the Jagiellonian University in Cracow are presented.

H. Wojtowicz (✉)
University of Rzeszow, Faculty of Mathematics and Nature,
Institute of Computer Science, Rzeszow, Poland
e-mail: hubert.wojtowicz@gmail.com

W. Wajs
AGH University of Science and Technology, Faculty of Electrical Engineering,
Institute of Automatics, Krakow, Poland
e-mail: wwa@ia.agh.edu.pl

B. Iantovics and R. Kountchev (eds.), *Advanced Intelligent Computational Technologies and Decision Support Systems*, Studies in Computational Intelligence 486, DOI: 10.1007/978-3-319-00467-9_1, © Springer International Publishing Switzerland 2014

1 Introduction

The diagnosis of Down's syndrome is not always easy, especially in the newborn infant. Clinical diagnosis of Down's syndrome is established by finding a set of typical small abnormalities localized in the area of face, palms and soles. The analysis of patterns of ridges and furrows of palms and soles (dermatoglyphs) may help in bringing the initial diagnosis. This analysis is carried out by an anthropologist. The method is non-invasive highly sensitive and specific but requires a lot of experience from a person carrying out the diagnosis. The critical stages or differentiation of dermal ridges occur in the third and fourth fetal months. Therefore, the dermatoglyphics are significant indicators of conditions existing several months prior to the birth of an individual. Dermatoglyphic patterns vary in relation to race and gender. A detection of abnormalities in the pattern distribution requires knowledge of pattern distributions for the general population. The distinction of the infants with Down's syndrome does not lie in the appearance of peculiar type of configurations, but it is rather an alteration of the frequencies of types which occur in healthy infants. Departures from the normal distribution of sets of patterns found in healthy infants are regarded as possible stigmata of genetic disorder. These distinctions are indicators of aberrant conditioning of dermal traits during the prenatal development. Diagnostic indexes were developed for several syndromes using predicative discrimination methods [16, 20]. The calculation of the indexes' scores allows reliable classification of infants into either normal or affected groups.

2 The Aim of the Project

The problem of pattern recognition and pattern understanding of genetic traits of infants with Down syndrome is a difficult and complex issue. The classification of genetic disorder on the basis of patterns is conducted by an anthropologist. A telemedical system of pattern recognition and pattern understanding can support anthropologist's work in the area of screening tests of data delivered from distant medical centers lacking specialist labs and personnel. The source of information in the form of patterns can originate in non-invasive way in any place equipped with a digital scanner and the Internet access. Transferred patterns can be analyzed in regard to features characteristic for Down syndrome and then classified.

In the project a following two stage information processing scheme is proposed:

1. A text analysis of specialist domain knowledge written in the form of natural language sentences, numerical tables, arithmetic expressions and arithmetical–logical relations. The analysis leads to the formulation of conditions from which the conclusion results.
2. A synthesis of digital information contained in the digitally stored pattern, which leads to the conditions generation from which the conclusion stems.

The basis of the scientific technique constitutes a system extracting features from the digital pattern of the image. The project makes use of image processing, pattern recognition and computational intelligence algorithms. Abnormalities of the patterns treated as features of the dermatoglyph can be found in the process of data extraction by the aforementioned algorithms. The system uses image pre-processing methods, which help to reduce the amount of data processed in the later stages by algorithms solving identification and classification problems. Basing on patterns of dermal ridges, and on the images containing information about the number of ridges and their directions it is possible to create classifiers recognizing traits of particular genetic disorders. For the detection of Down's syndrome the system uses rules of a diagnostic index called dermatoglyphic nomogram [9], which was developed for the screening test of infants suspected of having Down's syndrome [17].

3 Clinical Issues of Down's Syndrome

In the period of infancy a final diagnosis of Down's syndrome and congenital defects must be established. The diagnosis of defects, which because of its immediate threat to life requires an early surgical intervention is particularly important. TIt includes gastro-intestinal tract defects, which occur in 12 % of infants with Down's syndrome. The most common defects are duodenal atresia (2–5 %), Hirschprung's disease (2 %), tracheo esophageal fistula, hyperthropic pyloric stenosis, annular pancreas and anal and (or) rectal atresia. All these defects require an immediate surgical correction and according to the current rules infants with Down's syndrome must have the same standard of treatment as other patients. Other possible defects include congenital heart diseases, which occur in 40–50 % of infants with Down's syndrome. The most common abnormalities are endocardial cushion defect (atrioventricular canal), ventricular septal defect, tetralogy of Fallot and patent ductus arteriosus. Heart diseases require an early diagnosis because of the risk of congestive heart failure or pulmonary hypertension. An undiagnosed heart disease may lead to the impairment of physical development. Because of this risk every infant with Down's syndrome must be subjected to full diagnostics of circulatory system with respect to: ECG, chest radiography and echocardiography. Another serious defect that may occur is a cataract. Congenital cataract, which occurs in 3 % of infants with Down's syndrome, may significantly limit proper stimulation of retina. An early detection and removal of the lesion with subsequent correction by glasses may protect from significant visual impairment or blindness.

4 Description of Dermatoglyphic Nomogram

The dermatoglyphic nomogram is based upon four pattern areas and uses four dermatoglyphic traits chosen for their high discriminant value. These traits include: pattern types of the right and left index fingers, pattern type of the hallucal area of the right sole and value of the ATD angle of the right hand.

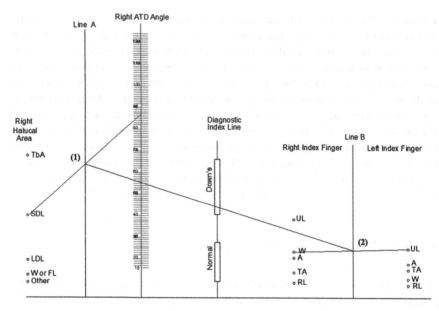

Fig. 1 Example of the dermatoglyphic nomogram. *TbA* tibial arch, *SDL* small distal loop, *LDL* large distal loop, *FL* fibular loop, *W* whorl, *UL* ulnar loop, *A* arch, *TA* tented arch, *RL* radial loop

Figure 1 presents an example of the nomogram. To determine a patient's index score three lines are constructed:

1. The first of the lines connects a particular determined type of pattern of the hallucal area of the sole with the selected value on the scale of the diagnostic line representing the determined value of the ATD angle of the right palm. In the case shown in the diagram it is the connection of the small distal loop pattern with the angular value equal to 85°. The plotted line intersects with line A at the point marked (1).

2. The second line connects the determined type of the pattern of the index fingerprint of the right hand with the pattern of the corresponding index fingerprint of the left hand. In this case it is the connection of the pattern type called whorl (W) with the pattern type called ulnar loop (UL). The plotted line intersects with line B at the point marked (2).

3. The last third line connects the point of intersection labelled as (1) with the point of intersection labelled as (2) and crosses the main diagnostic line. The point of intersection of the third line with the diagnostic line determines the diagnosis. There are three diagnostic cases, which correspond to the particular intervals of the diagnostic line. The first case corresponds to the intersection of the lines in the interval denoted as "Down's" an infant according to the determined value of the nomogram has Down's syndrome. The second case corresponds to the intersection of the lines in the interval denoted as "Normal" an infant according to the determined value of the nomogram is healthy. The

third case corresponds to the intersection of the lines in the interval between "Down's" and "Normal" it cannot be determined whether the infant has a genetic disorder or not.

Designations referring to the anatomy of the hand are used in describing palmar dermatoglyphics and in presenting methods of interpreting them. Terms of anatomical direction (proximal, distal, radial, ulnar) are employed in describing the locations of features and in indicating directions toward the respective palmar margins. Fingerprint pattern left loop is called ulnar loop when found on any of the fingers of the left hand and it is called radial loop when found on any of the fingers of the right hand. Pattern right loop in the left hand is called radial loop and in the right hand is called ulnar loop [4].

5 Literature Overview

Automatic systems exist for fingerprint classification. They are one of the tasks of dermatoglyphic analysis. These systems are not used in medical applications and don't take into account features appearing in dermatoglyphic impressions of the patients with genetic disorders. No research is carried out concerning problems like automatic counting of ridges between palmprint's singular points or automatic calculation of palmprint's ATD angle on the basis of singular points locations. No research is conducted in the area of automatic classification of patterns of the hallucal area of soles.

Support Vector Machine (SVM), applied to the classification of fingerprint patterns and hallucal area patterns, is a relatively new learning method which shows excellent performance in many pattern recognition applications. It maps an input sample into a high dimensional feature space and tries to find an optimal hyper plane that minimizes the recognition error for the training data by using the non-linear transformation function. SVM was originally designed for binary classification, and later extended to the multi-class problems.

Several systems for the classification of fingerprint impressions using SVM algorithms were described in the scientific literature. Min et al. [12] propose an approach combining supervised and non-supervised classification methods. A feature extracted in this approach from the images of fingerprint impressions is called FingerCode. FingerCode feature was proposed by Jain in (1999) [11]. FingerCode is extracted from the fingerprint image using a filter-based method. The algorithm sets a registration point in a given fingerprint image and tessellates it into 48 sectors. Then, it transforms the image using the Gabor filter of four directions (0°, 45°, 90°, and 135°). Ridges parallel to each filter direction are accentuated, and ridges not parallel to the directions are blurred. Standard deviations are computed on 48 sectors for each of the four transferred images in order to generate the 192-dimensional feature vector. Vectors created in this way for each of the fingerprint impressions are classified by the ensemble of one-vs-all

SVM algorithms. Values of one-vs-all decomposition scheme are stored in the decision templates. The decision templates (single-DTs) are generated for each of the classes by averaging the decision profiles (DPs) for the training samples. In the test stage the distance between the decision profile of a new sample and the decision templates of each class is computed. The class label is decided as the class of the most similar decision templates. In the described work a concept of single decision templates is extended to multiple decision templates. Multiple decision templates contain decision profiles generated from each of the OVA SVM ensemble classifiers. Multiple decision template contains all decision profiles for all of the five classes from each of the ensemble classifiers. The MuDTs decompose one class into several clusters to produce decision templates of each cluster. Decision profiles are clustered using self organizing maps algorithm. The classification process of the MuDTs is similar to that used with single-DTs. The distance between the decision profile of a new sample and each decision template of clusters is calculated, and then the sample is classified into the class that contains the most similar clusters. Euclidean distance is used to measure the similarity of decision profiles. The proposed method is tested on the NIST-4 database. From the fingerprint impressions contained in the database a FingerCode feature is extracted. Achieved classification accuracy on the test set is 90.4 %. Rejection ratio for the training set is 1.8 % and for the test set is 1.4 %.

In Hong and Cho [7] and Hong et al. [8] a method is proposed based on combination of outcomes from OVA SVM ensemble and naive Bayes classifier. In order to improve OVA SVM ensemble classification accuracy, all ensemble classifiers are dynamically ordered with naive Bayes classifiers on the basis of posterior probability of each class given the observed attribute values for a tested sample. The naive Bayes estimates the posterior probability of each class from the features extracted from the fingerprint images, which are singular points and pseudo ridges, while the OVA SVM ensemble is trained with FingerCode feature and generates probability margins for the classes. A subsumption architecture is used to cope with cases of ties and rejects in the process of voting fusion of OVA SVM ensemble. The evaluation order of the OVA SVMs is determined by the posterior probability of each class that the naive Bayes produces. The corresponding OVA SVM of a more probable class takes precedence in the subsumption architecture over the other OVA SVMs. The input sample is sequentially evaluated by each of the OVA SVM ensemble classifiers in order of decreasing probabilities calculated by naive Bayes classifier until an OVA SVM is satisfied. After the evaluation process, if the OVA SVM is satisfied, the sample is classified in the corresponding class of the OVA SVM. If none of the OVA SVM ensemble classifiers ascertains class membership of the sample it is classified into the class of the highest probability calculated by the naive Bayes classifier. Ordering the OVA SVMs properly for an input sample produces dynamic classification. Singular points used for the training of the naive Bayes classifier are detected by the Poincare index calculated from the orientation map of the ridges. Number and locations of fingerprint singular points unambiguously and uniquely determine fingerprint class to which the pattern of fingerprint impression belongs. Second of the analyzed features by the naive Bayes classifier is a PseudoCodes also known as PseudoRidges. PseudoRidges were proposed by Zhang and Yan [22] and they consist of a predefined number of points (100 points in the described work). A pseudo ridge is composed of 200 points based on orientation in two opposite directions from the starting point C. Proposed classification approach achieves classification accuracy of 90.8 % on the test set from the NIST-4 database.

In Min et al. [13] a comparison of approaches described in Min et al. [12] and Hong and Cho [7] was made. The first of the approaches uses multiple decision templates generated by OVA SVM ensemble of classifiers. The second approach uses naive Bayes classifier for the dynamic ordering of the OVA SVM ensemble classifiers. Both of the proposed approaches are tested on the NIST-4 database. The approach using multiple decision templates achieves classification accuracy of 90.8 % on the test set, while the approach using OVA SVM ensemble dynamically ordered witch the naive Bayes classifier produces classification accuracy of 90.4 % on the test set. In the chapter training and testing results are also presented for the OVA SVM ensemble, which achieves classification accuracy of 90.1 % on the test set and for naive Bayes classifier, which yields classification accuracy of 85.4 % on the test set.

In Ji and Yi [9] six class fingerprint classification scheme is used. The scheme used in the chapter classifies fingerprint patterns into the following classes: left loop (LL), right loop (RL), twin loop (TL), whorl (W), arch (A) and tented arch (TA). From the impression image gradients of local ridge directions are extracted and then in order to decrease calculations time these values are averaged to the four directions of values equal to (0, $\pi/4$, $\pi/2$ and $3\pi/4$). For each of these averaged directions a ratio of the number of the image segments which directions are equal to the particular direction to the number of all image segments is determined. Percentage ratios calculated for all of the four averaged directions create a vector classified by the OVA SVM ensemble. Experiment is conducted on the DB1 database from the FVC 2004 competition. The database contains 800 fingerprint impressions, 8 impressions for each of the fingerprints. Half of the data is used for the training of the OVA SVM ensemble, the other half is used in the testing phase of the classification. For the six class scheme achieved classification accuracy on the test set is 95.17 %.

In Li et al. [10] features of orientations of ridges and singular points locations are used to construct vectors which are classified by the SVM algorithm. In the proposed approach candidate singular points are extracted using filter based methods. These candidate singular points are then used to reconstruct the original orientation image using constrained nonlinear phase model of fingerprint orientation. The verification of the candidate singular points is achieved by computing the difference between the reconstructed orientation and the original orientation with high confidence level near the candidate singular points. The scheme in which orientation and singularities validate one another can be considered as an interactive validation mechanism. The best results for both the singular points and orientation image are selected to form the feature vectors for classification. Proposed feature extraction approach using SVM algorithms for the classification stage achieves 93.5 % classification accuracy on the test set coming from the NIST-4 database.

In all of the works basing on the NIST-4 database 17.5 % of the images are cross-referenced with two class labels due to the ambiguity of the pattern class, which is a result of simultaneous large intraclass variability and small interclass variability of the fingerprint patterns. The first label was only considered in training of the classifiers while in the testing phase outputs that corresponds to either the primary class or secondary class are taken as a correct classification. In our approach all of the fingerprint impressions in the database coming from the CMUJ are labeled with only a single label by an anthropologist and cross-labeling is not used in the testing phase of classification.

6 Classification Method of Fingerprint Patterns

Fingerprint classification is one of the tasks of dermatoglyphic analysis. Many classification methods were developed and described in the literature. The classification method used in dermatoglyphic analysis is called the Henry method. It classifies fingerprints into five distinct classes called: left loop (LL), right loop (RL), whorl (W), arch (A) or plain arch (PA) and tented arch (TA) (Fig. 2).

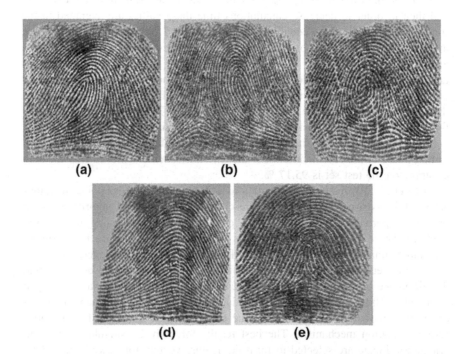

Fig. 2 Example fingerprints: **a** left loop; **b** right loop; **c** whorl; **d** plain arch; **e** tented arch

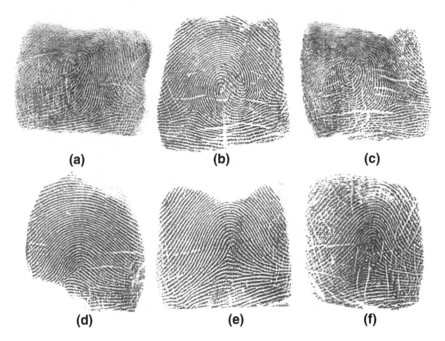

Fig. 3 The *upper row* contains patterns belonging to the same class but of different topology (large intraclass variability), the *lower row* contains patterns belonging to the different classes but of similar topology (small interclass variability)

The classification scheme based on the Henry method is a difficult pattern recognition problem due to the existence of small interclass variability of patterns belonging to the different classes and large intraclass variability of patterns belonging to the same class. The upper row in Fig. 3 shows three fingerprint impressions of different topology all belonging to the whorl class, the lower row in Fig. 3 shows from the left impressions of the fingerprints belonging to the classes: plain arch, tented arch and right loop, respectively.

In the chapter a classification scheme based on the extraction of fingerprint ridge orientation maps from the enhanced images has been presented. Vectors constructed from the directional images are used for training of SVM multi-class algorithms. For the induction process of SVMs we use RBF and triangular type kernels.

7 Classification Method of Patterns of the Hallucal Area of Sole

The classification of patterns in the hallucal area of soles is another task for dermatoglyphic analysis. The patterns in the hallucal area are classified into five distinct classes called: large distal loop (LDL), small distal loop (SDL), whorl (W), tibial arch (TA) and tibial loop (TL) (Fig. 4).

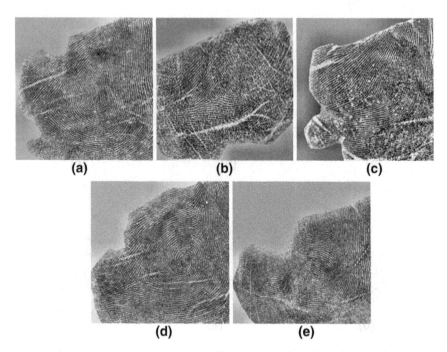

Fig. 4 Example patterns in the hallucal area of soles: **a** large distal loop, **b** small distal loop, **c** tibial arch, **d** whorl, **e** tibial loop

The classification of the patterns in the hallucal area of soles is also a difficult pattern recognition problem due to the existence of small interclass variability of patterns belonging to the different classes and large intraclass variability of patterns belonging to the same class. The upper row in Fig. 5 shows three impressions of different topology all belonging to the whorl class, the lower row in Fig. 5 shows on the left impression belonging to the class tibial arch, and on the right impression belonging to the class small distal loop.

8 The Method of ATD Angle Calculation

The hypothenar pattern of the palmprint creates an axial triradius (t) from which an ATD angle can be measured. The position of this axial triradius can be proximal (t), intermediate (t'), or distal (t"). The four triradii labeled A, B, C and D are located in a proximal relation to the bases of digits. The higher the axial triradius, the larger the ATD angle. Mean value of the ATD angle for normal infants is $\sim 48°$. The mean value of ATD angle for infants with Down's syndrome is $\sim 81°$.

The locations of palmprint's singular points are calculated from the orientation map. The orientation map is computed using algorithm based on the principal component analysis and multi-scale pyramid decomposition of the image.

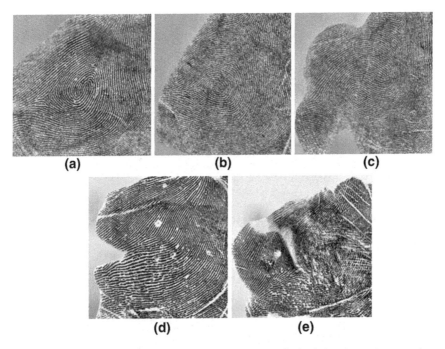

Fig. 5 The *upper row* contains patterns of the hallucal area of soles belonging to the same class but of different topology (large intraclass variability), the *lower row* contains patterns belonging to the different classes but of similar topology (small interclass variability)

Principal Component Analysis is applied to the image to calculate maximal likelihood estimations of the local ridge orientations and the multi-scale pyramid decomposition of the image helps to improve the accuracy of the calculated orientations. The algorithm calculating ridge orientations is robust to noise and allows for reliable estimation of local ridge orientations in the low quality areas of images. After the process of estimation of ridge orientations, Poincare index is calculated, which in turn allows for extraction of singular points from the palmprint image. Basing on the locations of singular points the ATD angle of the palmprint is calculated.

A two stage algorithm is applied in order to reliably estimate positions of singular points. In the first stage an improved formula of Poincare index is used to calculate candidate locations of singular points [21]. In the second stage a coherence map of the palmprint image is calculated. The coherence map is calculated for each pixel in the image on the basis of its eigenvalues. Pixel eigenvalues are calculated from the confusion matrix containing values calculated for each pixel (i, j) by applying the combination of two dimensional orthogonal Gaussian-Hermite moments to the segment of the image centered in pixel (i, j). The proposed approach combines information about singular points obtained from the improved Poincare index calculated from the orientation map and the

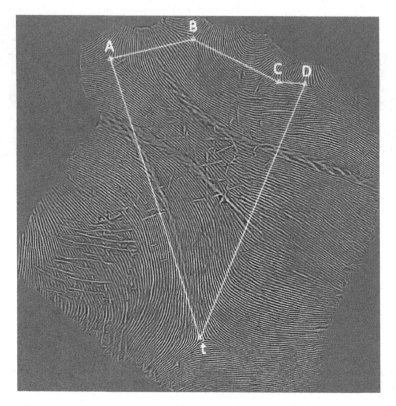

Fig. 6 Singular points of the palmprint located using two stage algorithm taking advantage of improved Poincare index and Gauss-Hermite moments

information obtained from the coherence map calculated from the pixel map of the image. The two stage approach allows for a reliable detection of only true singular points even in low quality areas of the palmprint image [21]. An example of calculation of palmprint's singular points using proposed algorithm is presented in Fig. 6.

9 Feature Extraction

Fingerprint impressions of low quality complicate the learning process of computational intelligence algorithms, and negatively influence their ability to accurately recognize the patterns. The classification accuracy can be improved by pre-processing of the images analyzed, which enhances their quality.

9.1 Image Preprocessing

Image pre-processing of fingerprint impressions consists of several stages. In the first stage image segmentation takes place, which separates background from the parts of the image where ridges are present. Segmentation algorithm calculates the histogram of the entire image and on that basis a threshold is selected. Areas of the image for which local histogram values are lower than the threshold are treated as the background [14]. After the region mask representing the foreground area is determined, the background area is removed from the image. Coordinates of the boundary points of the region mask are found. The image is truncated according to the boundary points coordinates and then using bicubic interpolation resized to the frame of 512×512 pixels. In the next stage image contrast enhancement is performed using CLAHE (Contrast Limited Adaptive Histogram Equalization) algorithm [23]. CLAHE divides the image into tiles. Each tile contrast is enhanced so that the resulting contrast histogram approximately corresponds to the shape of the statistical distribution specified as an input parameter for the CLAHE algorithm. Adjoining tiles are then combined using bilinear interpolation which smoothes inaccuracies on the edges of the tiles. The contrast in areas of homogeneous texture can be suppressed in order to prevent the amplification of noise that is always present in some form in the image of a fingerprint impression created using the offline ink method. In the last stage of pre-processing image quality enhancement is carried out using a contextual image filtration STFT (Short Time Fourier Transform) algorithm [3], which generates information about ridges flow directions, frequency of ridges and local image quality estimation.

9.2 Ridge Orientation

Ridge orientation maps are computed using algorithm based on Principal Component Analysis and multi-scale pyramid decomposition of the images [5]. Principal Component Analysis is applied to the fingerprint image to calculate the maximal likelihood estimations of the local ridge orientations. PCA based estimation method uses Singular Value Decomposition of the ensemble of gradient vectors in a local neighborhood to find the dominant orientation of the ridges. The algorithm calculating the ridge orientation maps also uses a multi-scale pyramid decomposition of the image, which helps to improve accuracy of the estimation.

The multi-scale method provides robustness to the noise present in the image. Examples of the ridge local orientations estimation in the distorted areas of the fingerprint impression from Fig. 7 are shown in Fig. 9a and b. Examples of the estimation of the local ridge orientations for the patterns of the hallucal area of sole from Fig. 8 are presented in Fig. 10a and b.

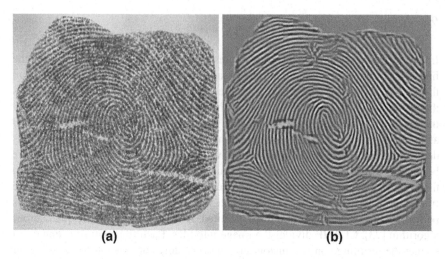

Fig. 7 An image of fingerprint impression belonging to the whorl class pre-processed using a contrast enhancement algorithm CLAHE (**a**), and then by filtration algorithm STFT (**b**)

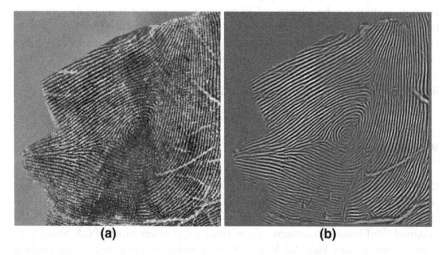

Fig. 8 An image of impression of the hallucal area belonging to the whorl class pre-processed using a contrast enhancement algorithm CLAHE (**a**), and then by filtration algorithm STFT (**b**)

10 Fingerprint Classification with SVM Algorithm

In this section SVM algorithm is presented and results of its application to the fingerprint classification task are discussed.

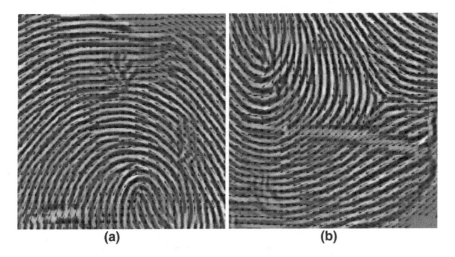

(a)   (b)

Fig. 9 Local values of ridge orientations in noised areas of fingerprint impression: **a** *top area* of the impression *above upper core* of the whorl, **b** *bottom left area* of the impression containing *left triradius* and *lower core* of the whorl

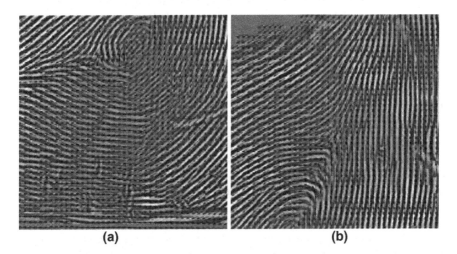

(a)   (b)

Fig. 10 Local values of ridge orientations in noised areas of the impression of the hallucal area of the sole: **a** *bottom area* of the impression below *lower core* of the whorl, **b** *top area* of the impression containing *upper triradius* and *upper core* of the whorl

10.1 Data

The data set consists of 600 fingerprint 8-bit grey scale images. The size of the images is 512 × 512 pixels. Classes left loop, right loop, whorl arch and tented arch are uniformly distributed in the data set and consist of 120 images each.

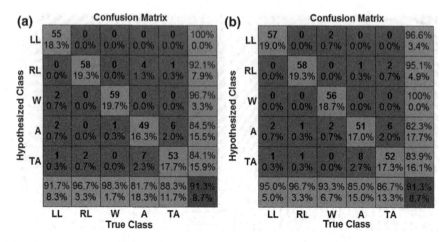

Fig. 11 Test results for the SVM algorithm trained with: **a** RBF kernel function, **b** triangular kernel function

A number of images containing partial or distorted information about the class of the pattern for which two or more image copies of the same impression were available were registered using non rigid registration algorithm and then mosaicked.

10.2 Classification Results

Fingerprint classification was accomplished using SVM algorithm. For multi-class problem an ensemble of SVM classifiers was created trained with one vs one voting method. Classifiers were using RBF type kernel functions and a triangular kernel [6]. The training dataset consists of 300 ridge orientation maps calculated for the fingerprint images. It contains 60 maps for each of the five classes. The dataset used for testing of the SVM is comprised of 300 ridge orientation maps. There are 60 maps of LL, RL, W, A and TA classes in the testing set. In the first stage images were pre-processed using the CLAHE algorithm and then filtered using STFT. Ridge orientation maps were computed from the filter-enhanced images using PCA and a multi-scale pyramid decomposition algorithm [5]. Cross-validation and grid search were used to obtain kernel parameters for the training of the SVM algorithms. Test results of the SVM algorithm with RBF and triangular kernel functions are presented as confusion matrices in Fig. 11a and b, respectively. Both the SVM trained with the RBF kernel and the SVM trained with the triangular kernel achieved classification accuracy of 91.3 % on the test data set.

11 Summary

In the present chapter a design of an automatic system for rapid screen diagnosing of trisomy 21 (Down's syndrome) in infants has been presented. Features extracted from the dermatoglyphic images are recognized by computational intelligence algorithms and results of these recognitions are passed as premises to the rules of the expert system. The design of the expert system is based on the dermatoglyphic nomogram. Tasks required for the determination of the patients' diagnostic score were described and methods of their implementation were discussed in the chapter. Results of dermatoglyphic patterns classification were presented in the chapter. Further research plan entails implementation of the classifier of the hallucal area of the sole patterns and implementation of the system for the calculation of the ATD angle of the palmprint. The final aim of the project is development and implementation of a comprehensive and widely available telemedical system supporting medical diagnosis.

References

1. Berg, C.: Harmonic analysis on semigroups: Theory of positive definite and related functions. Springer, Berlin (1984)
2. Boughorbel, S., Tarel, J.P.: Conditionally positive definite kernels for SVM based image recognition. IEEE International Conference on Multimedia and Expo (ICME), pp. 113–116 (2005)
3. Chikkerur, S., Cartwright, A.N., Govindaraju, V.: Fingerprint image enhancement using STFT analysis. Pattern Recogn. **40**, 198–211. Elsevier, Amsterdam (2007)
4. Cummins, H., Midlo, C.: Fingerprints, Palms and Soles—Introduction to Dermatoglyphics. Dover Publications Inc., New York (1961)
5. Feng, X.G., Milanfar, P.: Multiscale principal components analysis for image local orientation estimation. In: Proceedings of the 36th Asilomar Conference on Signals, Systems and Computers, vol. 1, pp. 478–482 (2002)
6. Fleuret, F., Sahbi, H.: Scale-invariance of support vector machines based on the triangular kernel. In: Proceedings of the IEEE International Conference on Computer Vision (ICCV) (2003)
7. Hong, J.H., Cho, S.B.: Dynamically subsumed-OVA SVMs for fingerprint classification. In: Proceedings of Pacific Rim International Conference on Artificial Intelligence (PRICAI), LNAI, vol. 4099, pp. 1196–1200. Springer, New York (2006)
8. Hong, J., Min, J., Cho, U., Cho, S.: Fingerprint classification using one-vs-all support vector machines dynamically ordered with naive Bayes classifiers. Pattern Recogn. **41**, 662-671. Elsevier, Amsterdam (2008)
9. Ji, L., Yi, Z.: SVM-based fingerprint classification using orientation field. In: Proceedings of Third International Conference on Natural Computation (ICNC), IEEE (2007)
10. Li, J., et al.: Combining singular points and orientation image information for fingerprint classification. Pattern Recogn. **41**, 353–366. Elsevier, Amsterdam (2008)
11. Maltoni, D., Maio, D., Jain, A.K., Prabhakar, S.: Handbook of Fingerprint Recognition. Springer, Berlin (2003)

12. Min, J.K., Hong, J.H., Cho, S.B.: Effective fingerprint classification by localized models of support vector machines. In: Proceedings of International Conference on Biometrics (ICB), LNCS vol. 3832, pp. 287–293. Springer, New York (2005)
13. Min, J.K., Hong, J.H., Cho, S.B.: Ensemble approaches of support vector machines for multiclass classification. In: Proceedings of Second International Conference on Pattern Recognition and Machine Intelligence (PReMI), LNCS vol. 4815, pp. 1–10. Springer, New York (2007)
14. Otsu, N.: A threshold selection method from gray-level histogram. IEEE Trans. Syst. Man Cybern. 9(1) 62–66 (1979)
15. Preuss, M.: A screening test for patients suspected of having Turner syndrome. Clin. Genet. 10, 145–155 (1976)
16. Reed, T.: Dermatoglyphics in Down's syndrome. Clin. Genet. (6), 236 (1974)
17. Reed, T.E., Borgaonkar, D.S., Conneally, P.M., Yu, P., Nance, W.E., Christian, J.C.: Dermatoglyphic nomogram for the diagnosis of Down's syndrome. J. Pediat. 77, 1024–1032 (1970)
18. Scholkopf, B.: The kernel trick for distances. In: Proceedings of Neural Information Processing Systems (NIPS), pp. 301–307 (2000)
19. Steinwart, I., Christmann, A.: Support Vector Machines. Springer Science (2008)
20. Tornjova-Randelova, S.G.: Dermatoglyphic characteristics of patients with Turner's syndrome. Med. Anthropol. 43(4), 96–100 (1990)
21. Yin, Y., Weng, D.: A new robust method of singular point detection from fingerprint. In: Proceedings of International Symposium on Information Science and Engineering, IEEE (2008)
22. Zhang, Q., Yan, H.: Fingerprint classification based on extraction and analysis of singularities and pseudo ridges. Pattern Recogn. 37(11), 2233–2243 (2004)
23. Zuiderveld, K.: Contrast Limited Adaptive Histogram Equalization. Graphics Gems IV. Academic Press (1994)

New Format for Coding of Single and Sequences of Medical Images

Roumen Kountchev, Vladimir Todorov and Roumiana Kountcheva

Abstract The recent development and use of huge image databases creates various problems concerning their efficient archiving and content protection. A wide variety of standards, methods and formats have been created, most of them aimed at the efficient compression of still images. Each standard and method has its specific advantages and demerits, and the best image compression solution is still to come. This chapter presents new format for archiving of still images and sequences of medical images, based on the Inverse Pyramid Decomposition, whose compression efficiency is comparable to that of JPEG. Main advantages of the new format are the comparatively low computational complexity and the ability to insert resistant and fragile watermarks in same digital image.

Keywords Image archiving · Lossy and lossless image compression · Image formats · Coding of medical image sequences

1 Introduction

The most famous contemporary file formats, widely used for image scanning, printing and representation, are TIF, JPG and GIF [1–3]. These are not the only choices of course, but they are good and reasonable choices for general purposes.

R. Kountchev (✉)
Department of Radio Communications and Video Technologies, Technical University—
Sofia, Bul. Kl. Ohridsky 8, 1000 Sofia, Bulgaria
e-mail: rkountch@tu-sofia.bg

V. Todorov · R. Kountcheva
T and K Engineering, Mladost 3, 1712 Sofia, Post Box 12, Bulgaria
e-mail: todorov_vl@yahoo.com

R. Kountcheva
e-mail: kountcheva_r@yahoo.com

B. Iantovics and R. Kountchev (eds.), *Advanced Intelligent Computational Technologies and Decision Support Systems*, Studies in Computational Intelligence 486, DOI: 10.1007/978-3-319-00467-9_2, © Springer International Publishing Switzerland 2014

Newer formats like JPG2000 never acquired popular usage, and are not supported by web browsers. The most popular format for medical images is DICOM (based on the JPEG standard) [4]. A brief survey of the most important contemporary image formats is given below.

The TIF format (Tag Image File Format) is the format of choice for archiving important images [5]. It is the most universal and widely supported format across all platforms, Mac, Windows, and UNIX. The main disadvantage is that TIF files are generally pretty large (uncompressed TIF files are about the same size in bytes as the image size in memory).

The JPG format (Joint Photographic Experts Group) [6] always uses lossy compression, but its degree is selectable: for higher quality and larger files, or lower quality and smaller files. However, this compression efficiency comes with a high price: some image quality is lost when the JPG data is compressed and saved, and this quality can never be recovered. This makes JPG be quite different from all the other usual file format choices. Even worse, more quality is lost every time the JPG file is compressed and saved again, so ever editing and saving a JPG image again is a questionable decision. JPG files are very small files for continuous tone photo images, but JPG is poor for graphics. JPG requires 24 bit color or 8 bit grayscale, and the JPG artifacts are most noticeable in the hard edges in the picture objects.

The GIF format (Graphic Interchange Format) [6] uses indexed color, which is limited to a palette of only 256 colors and is inappropriate for the contemporary 24-bit photo images. GIF is still an excellent format for graphics, and this is its purpose today, especially on the web. GIF uses lossless LZW compression and offers small file size, as compared to uncompressed data.

The PNG (Portable Network Graphics) format was designed recently, with the experience advantage of knowing all that went before [6]. It supports a large set of technical features, including superior lossless compression from LZ77 algorithm. The main disadvantage is that it incorporates special preprocessing filters that can greatly improve the lossless compression efficiency, but this introduces some changes in the processed image.

Depending on the format used, image compression may be lossy or lossless. Lossless compression is preferred for archival purposes and often for medical imaging, technical drawings, clip art, or comics. Lossy methods are especially suitable for natural images such as photographs in applications where minor (sometimes imperceptible) loss of fidelity is acceptable to achieve a substantial reduction in bit rate.

The basic methods for lossless image compression are:

- Run-length encoding—used as default method in PCX (Personal Computer eXchange—one of the oldest graphic file formats) and as one of possible in TGA (raster graphics file format, created by Truevision Inc.), BMP and TIFF;
- DPCM (Differential Pulse Code Modulation) and Predictive Coding;
- Adaptive dictionary algorithms such as LZW (Lempel–Ziv–Welch data compression)—used in GIF and TIFF;

- Deflation—used in PNG and TIFF
- Chain codes.

The most famous contemporary methods for lossy image compression are based on:

- Reducing the color space to the most common colors in the image;
- Chroma subsampling: this approach takes advantage of the fact that the human eye perceives spatial changes of brightness more sharply than those of color, by averaging or dropping some of the chrominance information in the image;
- Transform coding: a Fourier-related transform, such as DCT or the wavelet transform are applied, followed by quantization and entropy coding.
- Fractal compression: it differs from pixel-based compression schemes such as JPEG, GIF and MPEG since no pixels are saved. Once an image has been converted into fractal code, it can be recreated to fill any screen size without the loss of sharpness that occurs in conventional compression schemes.

The best image quality at a given bit-rate (or compression ratio) is the main goal of image compression, however, there are other important properties of image compression schemes:

Scalability generally refers to a quality reduction achieved by manipulation of the bitstream or file (without decompression and re-compression). There are several types of scalability:

- **Quality progressive** or layer progressive: The bitstream successively refines the reconstructed image.
- **Resolution progressive**: First encode a lower image resolution; then encode the difference to higher resolutions.
- **Component progressive**: First encode grey; then color.

Region of interest coding: Certain parts of the image are encoded with higher quality than others. This quality is of high importance, when medical visual information is concerned.

Meta information: Compressed data may contain information about the image which may be used to categorize, search, or browse images. Such information may include color and texture statistics, small preview images, and author or copyright information.

Processing power (computational complexity): Compression algorithms require different amounts of processing power to encode and decode.

All these standards and methods aim processing of single images. The contemporary medical practice involves also processing of large sequences of similar images, obtained from computer tomographs, and there is no special tool for their archiving.

In this chapter is presented one new format for still image compression and its extension for group coding of similar images. The chapter is arranged as follows: In Sect. 2 are given the basics of the methods, used for the creation of the new image archiving format; in Sect. 3 is presented the detailed description of the new format; in Sect. 4 are outlined the main application areas and Sect. 5—the Conclusion.

2 Brief Representation of the Methods, Used as a Basis for the Creation of the New Format

Two methods were used for the creation of the new format: the Inverse Pyramid decomposition (IDP) and the method for adaptive run-length data encoding (ARL).

2.1 Basic Principles of the IDP Decomposition

The IPD essence is presented in brief below for 8-bit grayscale images, as follows. The digital image matrix is first processed with two-dimensional (2D) direct Orthogonal Transform (OT) using limited number of low-frequency coefficients only. The values of these coefficients build the lowest level of the pyramid decomposition. The image is then restored, performing Inverse Orthogonal Transform (IOT) using the retained coefficients' values only. In result is obtained the first (coarse) approximation of the original image, which is then subtracted pixel by pixel from the original one. The so obtained difference image, which is of same size as the original, is divided into 4 sub-images and each is then processed with 2D OT again, using a pre-selected part of the transform coefficients again. The calculated values of the retained coefficients build the second pyramid level. The processing continues in similar way with the next, higher pyramid levels. The set of coefficients of the orthogonal transform, retained in every decomposition level, can be different and defines the restored image quality for the corresponding level (more coefficients naturally ensure higher quality). The image decomposition is stopped when the needed quality for the approximating image is obtained—usually earlier than the last possible pyramid level. The values of the coefficients got in result of the orthogonal transform from all pyramid levels are then quantitated, sorted in accordance with their spatial frequency, arranged as one-dimensional sequence, and losslessly compressed. For practical applications the decomposition is usually "truncated", i.e. it does not start from the lowest possible level but from some of the higher ones and for this, the discrete original image is initially divided into sub-blocks of size $2^n \times 2^n$. Each sub-block is then represented by an individual pyramid, whose elements are calculated using the corresponding recursive calculations. For decomposition efficiency enhancement is used a "truncated" decomposition, i.e. the decomposition starts after dividing the original image into smaller sub-images, and stops when the sub-image becomes of size 4×4 pixels.

The so presented approach was modified for the processing of a group of related images (usually these are images obtained by a computer tomography or in other similar cases). For this, one of the images is used as a reference one, and its first approximation for the lowest decomposition level is used by the remaining images in the group for the calculation of their next approximations. The detail presentation of the method and its applications are given in related publications of the authors [7, 8].

2.2 Description of the ARL Coding Method

The method for adaptive run-length data coding is aimed at efficient compression of sequences of same (or regularly changing) n-dimensional binary numbers without data length limitation. The coding is performed in two consecutive stages. In the first one, the input data is transformed without affecting their volume in such a way, that to obtain sequences of same values (in particular, zeros) of maximum length. The transform is reversible and is based on the analysis of the input data histograms and on the differences between each number in the processed data and the most frequent one. When the first stage is finished, is analyzed which data sequence is more suitable for the further processing (the original, or the transformed one) depending on the fact in which histogram were detected longer sequences of unused ("free") values. In case that the analyzed histograms do not have sequences, or even single free values, the input data is not suitable for compression; else, the processing continues with the second stage. In this part of the processing the transformed data is analyzed and sequences of same numbers are detected. Every such sequence is substituted by a shorter one, corresponding to the number of same values which it contains. Specific for the new method for adaptive run-length data coding is that it comprises operations of the kind "summing" and "sorting" only, and in result, its computational complexity is not high. The method is extremely suitable for compression of still images, which comprise mainly texts or graphics, such as (for example): bio-medical information (ECGs, EEGs, etc.), signatures, fingerprints, contour images, cartoons, and many others. The method is highly efficient and is suitable for real-time applications. The detailed presentation of the method is given in earlier publications of the authors [9, 10].

The new format, presented in detail below, comprises short descriptions of all parameters of the pyramid decomposition and of the lossless coding method, together with the basic parameters, needed for the group coding of similar images.

3 New Format Description

The format works with 8 and 24 bpp images of practically unlimited size. The compressed image data consists of two basic parts: the header and the data, obtained after the processing (i.e. the coded data). The header contains the information, which represents the values of the basic parameters of the IDP decomposition, needed for the proper decoding. It comprises 3 main parts: the IDP header, sub-headers for image brightness and color data, and information about the lossless coding of the transform coefficients' values. Additional information is added, needed for the processing of a group of similar images and compound images. The detailed description of the header follows below.

3.1 General Header

3.1.1 IDP Header

The first part of the header contains general information for the IDP decomposition configuration used:

Number of bytes after run-length coding (binary)—32 bits;
Final number of bytes, after entropy coding (binary)—32 bits;

IDP parameters—16 bits (truncated decomposition used, starting after image dividing into smaller sub-images):
Most significant byte (MSB):

MSB	X1	X2	X3	X4	X5	X6	X7	X8

X1–X4   Number of the selected initial decomposition level (1 hexadecimal number);
X5–X8   Number of the selected end decomposition level (1 hexadecimal number);

Least significant byte (LSB):

LSB	X9	X10	X11	X12	X13	X14	X15	X16

X9–X12   Maximum number of pyramid levels (1 hexadecimal number)—depends on the processed image size;
X13   Direct coding (optional);
X14   Free;
X15   Color/Grayscale;
X16   Color palette type

Description of the basic decomposition parameters—16 bits:

MSB	X1	X2	X3	X4	X5	X6	X7	X8

X1   Gray type color image;
X2   Inverse Pyramid decomposition;

X3   Difference/Direct image coding;
X4, 5   Free/Lossless compression;
X6, 7   Color space used (PAL, NTSC, RCT, YCrCb);
Coding   00—PAL; 01—NTSC; 02—RCT; 03—YCrCb;
X8   Color space RGB selected

LSB	X9	X10	X11	X12	X13	X14	X15	X16

X9   Horizontal image scan;
X10 Run-length coding performed in horizontal direction;
X11 Right shift for brightness components data (quantization);
X12 Right shift for color components data (quantization);
X13 Entropy coding enabled;
X14 Run-length encoding enabled;
X15 Flag indicating that the U values in the YUV color transform had been used;
X16 Flag indicating that the V values in the YUV color transform had been used

3.1.2 Sub-Header for the Image Brightness Data

Mask for the transform coefficients in the initial pyramid level (4 hexadecimal numbers)—16 bits:

MSByte	MSb	....	LSb		LSByte	MSb	....	LSb

Common mask for the transform coefficients in the middle levels of the pyramid decomposition (4 hexadecimal numbers)—16 bits:

MSByte	MSb	.....	LSb		LSByte	MSb	.....	LSb

Mask for the transform coefficients in the highest pyramid level (4 hexadecimal numbers)—16 bits:

MSByte	MSb	.....	LSb		LSByte	MSb	.....	LSb

Run-length type, 8 types—8 bits (patented);
First Run-length code—8 bits (patented);
Last (second) Run-length code—8 bits (patented);
Quantization coefficient for the first pyramid level—binary number (8 bits);
Selected approximation method for each level (WHT, DCT, etc.)—6 bits: 2 bits for each group: Initial level, Middle levels and Last level. Approximation coding: 00—WHT; 01—DCT; 02—Plane; 03—Surface.

Original image size (vertical direction)—binary number (16 bits);
Original image size (horizontal direction)—binary number (16 bits).

3.1.3 Sub-Header for the Image Color Data

Number of the selected initial level (2 hexadecimal numbers)—8 bits:

Number of the selected end level (2 hexadecimal numbers)—8 bits:

Mask for the used coefficients selection in the initial decomposition level (4 hexadecimal numbers)—16 bits:

MSByte	MSb	....	LSb		LSByte	MSb	....	LSb

Mask for the used coefficients selection in the middle decomposition levels (common mask for all levels)—16 bits:

MSByte	MSb	....	LSb		LSByte	MSb	....	LSb

Mask for the used coefficients selection in the highest decomposition level—16 bits:

MSByte	MSb	....	LSb		LSByte	MSb	....	LSb

- Division coefficient for the first pyramid level (binary number)—8 bits;
- Selected approximation method for each level (WHT, DCT, etc.)—6 bits, 2 bits for each group: Initial level, Middle levels and Last level.

  Coding: 00—WHT; 01—DCT; 02—Plane; 03—Surface;

- Original image size (vertical direction)—binary, 16 bits;
- Original image size (horizontal direction) – binary, 16 bits;
- Used color format (4:1:1, 4:2:2, 4:4:4, 4:2:0)—16 bits.

  Coding: 00—4:2:0; 01—4:4:4; 02—4:2:2; 04—4:1:1.

3.1.4 Sub-Headers for Transform Coefficients Quantization

- Sub-header for the last level (brightness): 16 binary numbers, for coefficients (0–15): 8 bits each;

- Sub-header for the last level (color): 16 binary numbers, for coefficients (0–15): 8 bits each.

3.1.5 Entropy Coding Table

- Number of bytes representing the Entropy Table (8 bits);
- Coding tree:
  - most frequent value;
  - second most frequent value;
  - next values (up to 19 or 31), depending on the coding tree length permitted (selected after better performance evaluation).
  - Coded image data.

3.2 Coding of Group of Images: Additional Header

3.2.1 Group Header (First IDP Header)

Group coding identifier—16 bits (binary number);
Number of images in the group—8 bits (binary number);
Length of the reference image name—8 bits;
String of length equal to reference image name;
Length of the processed image name;
String of length equal to processed image name.

3.2.2 Header of the First Approximation of the Reference Image (Second Header)

- IDP header—contains the description of the predefined decomposition parameters.
- Coded data

3.2.3 Next Header (for Higher Level Approximations of Each Multispectral Image in the Group)

- Regular header—IDP decomposition parameters.
- Coded data

3.3 Coding of Compound Images: Additional Header

The basis is the IDP header—i.e. the first header.
   Additional information placed at the header end:

- Layer file identifier—16 bits (binary number)—layers correspond to number of objects detected in the image: pictures, text, graphics, etc.

   For each part (object):

- Coordinates of the image part segmented (in accordance with the object segmentation—text or picture)—2 × 16 bits (horizontal and vertical coordinate of the lower left corner of the image part);
- Corresponding regular IDP header;
- Coded data.

3.4 Additional Information

The new format is very flexible and permits the addition of meta information, which to enhance image categorization, search, or browsing: special data about the disease, the patient age, medication, etc. This could also include the kind of IDP watermarking used for the image content protection (resistant or fragile), the number of watermarks, or other similar information.

4 Application Areas

The everyday IT practices require intelligent approach in image compression and transfer. Usually such applications as image archives, e-commerce, m-trade, B2B, B2C, etc., need fast initial image transfer without many details, which are later sent to customers on request only. For such applications the IDP method is extremely suitable because it permits layered image transfer with increasing visual quality. Additional advantage of the IDP method is that because of the layered decomposition it permits insertion of multiple digital watermarks (resistant and fragile) in each layer [11]. Special attention requires the ability of the method for scalable image representation. For this, the image approximations, which correspond to lower decomposition layers, are represented scaled down. In result, the quality of the restored images is visually lossless for very high compression ratios [12]. Additional advantage is that the specially developed format permits easy insertion of meta-information of various kinds. The computational complexity of the method is lower than that of other contemporary methods for image compression. Detailed evaluation of the computational complexity and comparison with JPEG2000 is given in [12]. One more advantage is, that unlike JPEG, in which more quality is lost every time the file is compressed and saved again, the IDP format retains the restored image quality regardless of the number of compressions/decompressions. The method is suitable for wide variety of applications: still image compression and archiving; creation of contemporary image databases with reliable content protection; efficient compression of multispectral and multi-view images, medical images and sequences of medical images.

The flexible IDP-based image processing, used for the creation of the new format, permits setting of regions of interest and building of individual decomposition pyramids for each region of interest. This approach is the basis for adaptive processing of compound images, which contain pictures and texts.

5 Conclusions

The new format, presented in this chapter, is already implemented in software (Visual C, Windows environment) developed at the Technical University of Sofia. Several versions are already developed for still image compression, image group coding, image content protection with multiple watermarks, image hiding, archiving of multi-view and multispectral images and video sequences. The software implementation of the IDP method based on the new format confirms its flexibility and suitability for various application areas. The new format answers the requirements for the most important properties of the image compression schemes: Scalability, Region of interest coding, Processing power and Meta information. Compared to basic image compression standards, the new format offers certain advantages, concerning efficiency, computational complexity, ability to set regions of interest, and the insertion of meta information (setting regions of interest after image content analysis, digital watermarks insertion, etc.) [11]. The relatively low computational complexity permits its ability for real-time applications also.

Special advantage is the method version for lossless image and data compression, which is extremely efficient for graphics (EEGs and ECGs also) and text images [10].

Acknowledgments This chapter was supported by the Joint Research Project Bulgaria-Romania (2010–2012): "Electronic Health Records for the Next Generation Medical Decision Support in Romanian and Bulgarian National Healthcare Systems", DNTS 02/09.

References

1. Huang, H., Taira, R.: Infrastructure design of a picture archiving and communication system. Am. J. Roentgenol. **158**, 743–749 (1992)
2. Taubman, D., Marcellin, M.: JPEG 2000: Image Compression Fundamentals, Standards and Practice. Kluwer Academic Publishers, Boston (2002)

3. Information Technology–JPEG 2000 Image Coding System: Part 9–Interactivity tools, APIs and protocols, no. 15444-9, ISO/IEC JTC1/SC29/WG1 IS, Rev. 3 (2003)
4. Pianykh, O.: Digital Imaging and Communications in Medicine (DICOM). Springer, Berlin (2008)
5. TIFF—Revision 6.0. Adobe Developers Association (1992) http://www.adobe.com/Support/TechNotes.html
6. Miano, J.: Compressed Image File Formats: JPEG, PNG, GIF, XBM, BMP. Addison Wesley Professional, Reading (1999)
7. Kountchev, R., Kountcheva, R.: Image representation with reduced spectrum pyramid. In: Tsihrintzis, G., Virvou, M., Howlett, R., Jain, L. (eds.) New Directions in Intelligent Interactive Multimedia, pp. 275–284. Springer, Berlin (2008)
8. Kountchev, R., Kountcheva, R.: Compression of multispectral and multi-view images with inverse pyramid decomposition. Int. J. Reasoning-based Intell. Syst. 3(2), 124–131 (2011)
9. Kountchev, R., Todorov, V.L., Kountcheva, R.: New method for lossless data compression based on adaptive run-length coding. In: Enachescu, C., Filip, F., Iantovics, B. (eds.) Advanced Computational Technologies, Romanian Academy Publishing House, (in press)
10. Kountchev, R., Todorov, V.L., Kountcheva, R., Milanova, M.: Lossless compression of biometric image data. Proceedings of 5th WSEAS International Conference on Signal Processing, pp. 185–190. Istanbul, Turkey (2006)
11. Kountchev, R., Todorov, V.L., Kountcheva, R.: Fragile and resistant image watermarking based on inverse difference pyramid decomposition. WSEAS Trans. Sig. Process. 6(3), 101–112 (2010)
12. Kountchev, R., Milanova, M., Ford, C., Kountcheva, R.: Multi-layer image transmission with inverse pyramidal decomposition. In: Halgamuge, S., Wang, L. (eds.) Computational intelligence for modeling and predictions, Ch. 13, vol. 2, pp. 179–196. Springer, Berlin (2005)

Rule-Based Classification of Patients Screened with the MMPI Test in the Copernicus System

Daniel Jachyra, Jerzy Gomuła and Krzysztof Pancerz

Abstract The Copernicus system is a tool for computer-aided diagnosis of mental disorders based on personality inventories. Knowledge representation in the form of rules is the closest method to human activity and reasoning, among others, in making a medical diagnosis. Therefore, in the Copernicus system, rule-based classification of patients screened with the MMPI test is one of the most important parts of the tool. The main goal of the chapter is to give more precise view of this part of the developed tool.

1 Introduction

Computer support systems supporting medical diagnosis have become increasingly more popular worldwide (cf. [14]). Therefore, the tool called Copernicus [11] for classification of patients with mental disorders screened with the MMPI test has been developed. The Minnesota Multiphasic Personality Inventory (MMPI) test (cf. [6, 15]) delivering psychometric data on patients with selected mental

D. Jachyra (✉)
Chair of Information Systems Applications, University of Information Technology
and Management in Rzeszów, Rzeszów, Poland
e-mail: djachyra@wsiz.rzeszow.pl

J. Gomuła
The Andropause Institute, Medan Foundation, Warsaw, Poland
e-mail: jerzy.gomula@wp.pl

J. Gomuła
Cardinal Stefan Wyszyński University in Warsaw, Warsaw, Poland

K. Pancerz
Institute of Biomedical Informatics, University of Information Technology
and Management in Rzeszów, Rzeszów, Poland
e-mail: kpancerz@wsiz.rzeszow.pl

B. Iantovics and R. Kountchev (eds.), *Advanced Intelligent Computational Technologies and Decision Support Systems*, Studies in Computational Intelligence 486, DOI: 10.1007/978-3-319-00467-9_3, © Springer International Publishing Switzerland 2014

disorders is one of the most frequently used personality tests in clinical mental health as well as psychopathology (mental and behavioral disorders). In years 1998–1999, a team of researchers, consisting of W. Duch, T. Kucharski, J. Gomuła, R. Adamczak, created two independent rule systems, devised for the nosological diagnosis of persons, that may be screened with the MMPI-WISKAD test [7]. The MMPI-WISKAD personality inventory is a Polish adaptation of the American inventory (see [3, 16]). The Copernicus system developed by us is the continuation and expansion of that research.

In the Copernicus system, different quantitative groups of methods supporting differential inter-profile diagnosis have been selected and implemented. However, rule-based classification of patients is one of the most important parts of the tool. Knowledge representation in the form of rules is the closest method to human activity and reasoning, among others, in making a medical diagnosis. In the most generic format, medical diagnosis rules are conditional statements of the form:

IF *conditions*(*symptoms*), THEN *decision*(*diagnosis*).

The rule expresses the relationship between symptoms determined on the basis of examination and diagnosis which should be taken for these symptoms before the treatment. In our case, symptoms are determined on the basis of results of a patient's examination using the MMPI test. In the Copernicus system, the number of rule sets generated by different data mining and machine learning algorithms (for example: the RSES system [1], the WEKA system [17]) is included. However, the rule base can be extended to new rule sets delivered by the user. In the case of multiple sources of rules, we obtain a combination of classifiers. In classifier combining, predictions of classifiers should be aggregated into a single prediction in order to improve the classification quality. The Copernicus system delivers a number of aggregation functions described in the remaining part of this chapter.

The Copernicus system supports the idea that visualization plays an important role in professional decision support. Some pictures often represent data better than expressions or numbers. Visualization is very important in dedicated and specialized software tools used in different (e.g., medical) communities. In the Copernicus system, a special attention has been paid to the visualization of analysis of MMPI data for making a diagnosis decision easier. A unique visualization of classification rules in the form of stripes put on profiles as well as visualization of results of aggregated classification have been designed and implemented.

2 MMPI Data

In the case of the MMPI test, each case (patient) x is described by a data vector $a(x)$ consisting of thirteen descriptive attributes: $a(x) = [a_1(x), a_2(x), \ldots, a_{13}(x)]$. If we have training data, then to each case x we additionally add one decision attribute d—a class to which a patient is classified.

The validity part of the profile consists of three scales: L (laying), F (atypical and deviational answers), K (self defensive mechanisms). The clinical part of the profile consists of ten scales: 1.Hp (Hypochondriasis), 2.D (Depression), 3.Hy (Hysteria), 4.Ps (Psychopathic Deviate), 5.Mf (Masculinity/Femininity), 6.Pa (Paranoia), 7.Pt (Psychasthenia), 8.Sc (Schizophrenia), 9.Ma (Hypomania), 0.It (Social introversion). The clinical scales have numbers attributed so that a profile can be encoded to avoid negative connotations connected with the names of scales. Values of attributes are expressed by the so-called T-scores. The T-scores scale, which is traditionally attributed to MMPI, represents the following parameters: offset ranging from 0 to 100 T-scores, average equal to 50 T-scores, standard deviation equal to 10 T-scores.

For our research, we have obtained input data which has nineteen nosological classes and the reference class (*norm*) assigned to patients by specialists. Each class corresponds to one of psychiatric nosological types: neurosis (*neur*), psychopathy (*psych*), organic (*org*), schizophrenia (*schiz*), delusion syndrome (*del.s*), reactive psychosis (*re.psy*), paranoia (*paran*), sub-manic state (*man.st*), criminality (*crim*), alcoholism (*alcoh*), drug addiction (*drug*), simulation (*simu*), dissimulation (*dissimu*), and six deviational answering styles (*dev1, dev2, dev3, dev4, dev5, dev6*). The data set examined in the Copernicus system was collected by T. Kucharski and J. Gomuła from the Psychological Outpatient Clinic.

Data vectors can be represented in a graphical form as the so-called MMPI profiles. The profile always has a fixed and invariable order of its constituents (attributes, scales). Let a patient x be described by the data vector:

$$a(x) = [56, 78, 55, 60, 59, 54, 67, 52, 77, 56, 60, 68, 63].$$

Its profile is shown in Fig. 1.

A basic profile can be extended by additional indexes or systems of indexes (cf. [13]). Different combinations of scales constitute diagnostically important indexes (e.g., Gough's, Goldberg's, Watson-Thomas's, L'Abate's, Lovell's indexes—see [6]) and systems of indexes (e.g., Diamond's [5], Leary's, Eichmann's, Petersen's, Taulbee-Sisson's, Butcher's [2], Pancheri's). They have been determined on the basis of clinical and statistical analysis of many patients' profiles. All mentioned and some additional (e.g., the nosological difference-configuration Gough-Płużek's system) indexes have been implemented in the Copernicus system. It enables the user to extend the basic profile even to 100 attributes (13 scales plus 87 indexes).

Table 1 An input data for Copernicus (fragment)

Attribute	a_1	a_2	a_3	a_4	a_5	a_6	a_7	a_8	a_9	a_{10}	a_{11}	a_{12}	a_{13}	class
Scale	L	F	K	1.Hp	2.D	3.Hy	4.Ps	5.Mf	6.Pa	7.Pt	8.Sc	9.Ma	0.It	
#1	55	65	50	52	65	57	63	56	61	61	60	51	59	*norm*
#2	50	73	53	56	73	63	53	61	53	60	69	45	61	*org*
#3	56	78	55	60	59	54	67	52	77	56	60	68	63	*paran*
...	...	...	...	...	...	...	...	...	...	...	...	...	...	...

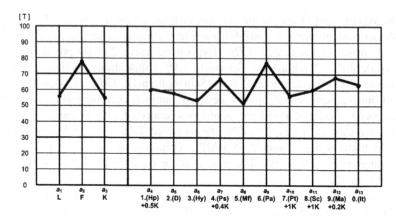

Fig. 1 MMPI profile of a patient (example); suppressors +0.5K, +0.4K, +1K, +0.2K—a correction value from raw results of scale K added to raw results of selected clinical scales

3 Rule-Based Classification in Copernicus

In this section, we describe step-by-step functionality of the Copernicus system concerning rule-based classification of patients screened with the MMPI test.

3.1 General Selection of Rule Sets

For classification purposes, the user can select a number of rule sets included in the tool (see Fig. 2). Such rule sets have been generated by different data mining and machine learning algorithms (for example: the RSES system [1], the WEKA system [17]). Moreover, the user can select its own rule set. The problem of selecting suitable sets of rules for classification of MMPI profiles has been considered in our previous chapters (see [8–10, 12, 13]).

3.2 Specific Selection of Rules

After general selection of rule sets, the user can determine more precisely which rules will be used in the classification process.

Each rule R in the Copernicus system has the form:

$$\text{IF } a_{i1}(x) \in [x_{i1}^l, x_{i1}^r] \text{ AND } \ldots \text{ AND } a_{ik}(x) \in [x_{ik}^l, x_{ik}^r], \text{ THEN } d(x) = d_m, \quad (1)$$

where $a_{i1}, \ldots, a_{ik}$ are selected scales (validity and clinical) or specialized indexes, $x_{i1}^l, x_{i1}^r, \ldots, x_{ik}^l, x_{ik}^r$ are the left and right endpoints of intervals, respectively, d is a

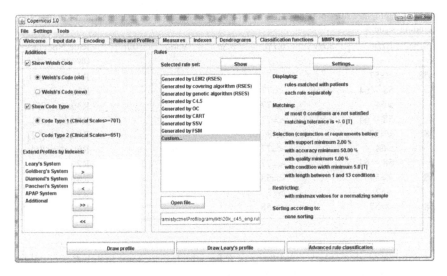

Fig. 2 The rule selection window in the Copernicus system

diagnosis, d_m is one of nosological classes proposed for the diagnosis. Each $a_{ik}(x) \in [x_{ik}^l, x_{ik}^r]$ is called an elementary condition of R. For each patient x, its profile is said to be matched to a rule R if and only if $a_{i1}(x) \in [x_{i1}^l, x_{i1}^r]$ and ... and $a_{ik}(x) \in [x_{ik}^l, x_{ik}^r]$. This fact is denoted by $x| = R$.

Each rule set included in the Copernicus system has been extracted from a proper set of cases called a training set. Each rule R in the form of 3.1 can be characterized by the following factors:

- the accuracy factor $acc(R) = \frac{n_C}{n_{CD}}$,
- the total support factor $supp_t(R) = \frac{n_C}{n}$,
- the class support factor $supp_c(R) = \frac{n_C}{n_D}$,
- the quality factor $qual(R) = acc(R)supp(R)$,
- the length factor $length(R) = card(a_{i1}, ..., a_{ik})$, i.e., a number of elementary conditions of R,

where $card$ denotes the cardinality of a given set, n is the size (a number of cases) of the training set, n_C is a number of cases in the training set which are matched to R, n_{CD} is a number of cases in the training set which are matched to R and which have additionally assigned the class d_m, n_D is a number of cases in the training set which have assigned the class d_m.

Exemplary classification rules obtained from data consisting of profiles using the WEKA system are shown in Table 2. The accuracy factor has been assigned to each rule.

The user can set, among others, that each rule R used for classification a given case x satisfies the following conditions:

Table 2 Exemplary rules obtained after transformation of a decision tree generated for all scales excluding scale 5

L	1.Hp	3.Hy	6.Pa	8.Sc	9.Ma	0.It	Rule No.:class (accuracy %)
<=58	<=57						R1:*norm* (89)
	58 – 64	<=59	<=77	<=68			R2:*norm* (89
	>64		<=77		<=56		R3:*neur* (85)
	58 – 64	>59	<=77	<=68			R4:*neur* (85)
>58	<=57		>58				R5:*psych* (92)
	>64		<=77		>56		R6:*schiz* (85)
	58 – 64		<=77	>68			R7:*schiz* (85)
>58	<=57		<=58			>59	R8:*simu* (94)
	>57		>77				R9:*simu* (94)
>58	<=57		<=58			<=59	R10:*dissimu*(85)

- a number of elementary conditions of R for which x does not match R is equal to 0 (exact matching) or more (approximate matching),
- x is matched to R with a certain degree (tolerance), i.e., for each elementary condition $a_{ik}(x) \in [x^l_{ik} - t, x^r_{ik} + t]$, where t is a tolerance value,
- $acc(R)$ is greater or equal to a given threshold,
- $supp(R)$ is greater or equal to a given threshold,
- $qual(R)$ is greater or equal to a given threshold,
- the interval $[x^l_{ik}, x^r_{ik}]$ of each elementary condition of R has the property that $x^r_{ik} - x^r_{ik}$ is greater or equal to a given threshold (i.e., a rule with tight conditions can be omitted),
- $length(R)$ is included in a given interval (i.e., a rule with too short or too long condition part can be omitted),
- indicated scales in elementary conditions of R are omitted—the so called scale excluding (for example diagnosticians' experience shows that the scale 5.*Mf* is weak and it should be omitted).

For elementary conditions in the form of intervals, we sometimes obtain lower and upper bounds that are, for example, $-\infty$ and ∞, respectively. Such values cannot be rationally interpreted from the clinical point of view. Therefore, ranges of classification rule conditions can be restricted. We can replace ∞ by:

- a maximal value of a given scale occurring for a given class in our sample,
- a maximal value of a given scale for all twenty classes,
- a maximal value of a given scale for a normalizing group (i.e., a group of women, for which norms of validity and clinical scales have been determined),
- a maximal value for all scales for a normalizing group, i.e., 120 T-scores.

A procedure for restricting ranges of classification rule conditions with the value $-\infty$ is carried out similarly, but we take into consideration minimal values. A minimal value for all scales of normalizing group of women is 28 T-scores. Specialized indexes are linear combinations of scales. Therefore, restricting ranges of rule conditions for indexes is also possible and simple. Copernicus enables us to select the way of restricting ranges. Rule conditions are automatically restricted to the form readable for the diagnostician-clinician.

3.3 Visualization of Profiles and Rules

The rule R can be graphically presented as a set of stripes placed in the profile space. Each condition part $a_{ij}(x) \in [x_{ij}^l, x_{ij}^r]$, where $j = 1, ..., k$, of the rule R is represented as a vertical stripe on the line corresponding to the scale a_{ij}. This stripe is restricted from both the bottom and the top by values x_{ij}^l, x_{ij}^r, respectively. Such visualization enables the user to easily determine which rule matches a given profile (cf. Fig. 3).

3.4 Classification Results

On the basis of rules a proper diagnostic decision for the case x can be made. Aggregation factors implemented in the Copernicus system enable selecting only one main decision from decisions provided by rules used for the classification of x. For each class d for which cases can be classified, a number of different aggregation factors can be calculated. The first aggregation factor is the simplest one.

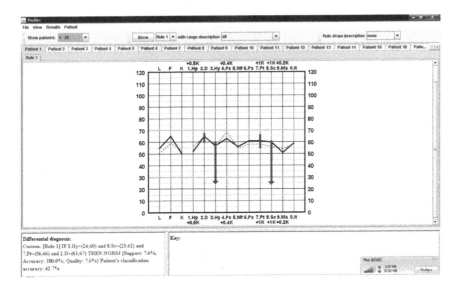

Fig. 3 Visualization of a rule in the profile space

It expresses the relative number of rules denoting the class d in the set of all rules matched by x in the selected sense (see Sect. 3.2):

$$aggr_1(d) = \frac{card(\{R : x| = R \text{ and } class(R) = d\})}{card(\{R : x| = R\})}.$$

Another three aggregation factors take also into consideration the maximal value of quality factors of rules from indicating the class d. These factors differ on the weights of components:

$$aggr_2(d) = 0.8 \max(\{qual(R) : x| = R \text{ and } class(R) = d\})$$
$$+ 0.2 \frac{card(\{R : x| = R \text{ and } class(R) = d\})}{card(\{R : x| = R\})},$$

$$aggr_3(d) = 0.5 \max(\{qual(R) : x| = R \text{ and } class(R) = d\})$$
$$+ 0.5 \frac{card(\{R : x| = R \text{ and } class(R) = d\})}{card(\{R : x| = R\})},$$

$$aggr_4(d) = 0.67 \max(\{qual(R) : x| = R \text{ and } class(R) = d\})$$
$$+ 0.33 \frac{card(\{R : x| = R \text{ and } class(R) = d\})}{card(\{R : x| = R\})}.$$

The last two aggregation factors additionally take into consideration the average length of rules from the d class. In this case, the smaller the average length is, the better the set of rules. These factors differ on the weights of components:

$$aggr_5(d) = 0.6 \max(\{qual(R) : x| = R \text{ and } class(R) = d\})$$
$$+ 0.2avg(\{1 - length(R) : x| = R \text{ and } class(R) = d\})$$
$$+ 0.2 \frac{card(\{R : x| = R \text{ and } class(R) = d\})}{card(\{R : x| = R\})},$$

$$aggr_6(d) = 0.4 \max(\{qual(R) : x| = R \text{ and } class(R) = d\})$$
$$+ 0.4avg(\{1 - length(R) : x| = R \text{ and } class(R) = d\})$$
$$+ 0.2 \frac{card(\{R : x| = R \text{ and } class(R) = d\})}{card(\{R : x| = R\})}.$$

In formulas of aggregation factors, *max* denotes the maximum value, *avg* denotes the arithmetic average value, and $class(R)$ denotes the class indicated by the rule R. To calculate the quality factor of a rule, we can use either the total support factor or the class support factor.

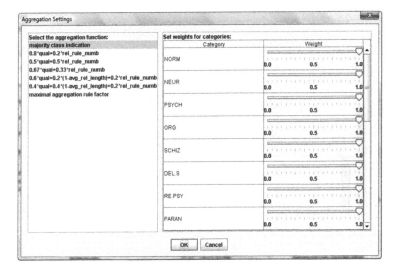

Fig. 4 Selection of aggregation factors in the Copernicus system

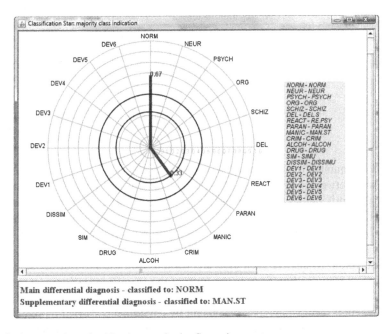

Fig. 5 An exemplary classification star in the Copernicus system

If a given aggregation factor $aggr(d)$ is calculated for each class d to which cases can be classified, then weighted maximum value is determined:

$$\max(\{w_1 aggr(d_1), w_2 aggr(d_2), \ldots, w_m aggr(d_m)\}),$$

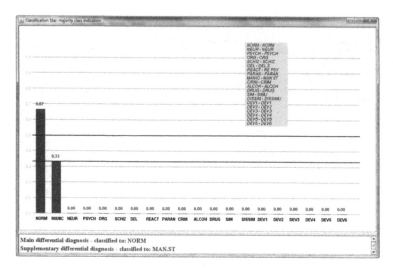

Fig. 6 An exemplary classification column chart in the Copernicus system

Aggregation factor	Quality of classification (%)
Table 3 A quality of classification of cases for described aggregation factors	
$aggr_1$	87.9
$aggr_2$	77.4
$aggr_3$	84.8
$aggr_4$	82.9
$aggr_5$	78.9
$aggr_6$	82.6

where m is a number of all possible classes and weights $(w_1, w_2, \ldots, w_m)$ can be set by the user between 0 and 1, for each class separately (see Fig. 4).

The main differential diagnosis for a given case x is set as d_m if $w_m aggr(d_m)$ is the maximum value and $w_m aggr(d_m) > 0.67$. The supplementary differential diagnosis for a given case x is set as d_s if $w_s aggr(d_s)$ is the next maximum value and $w_s aggr(d_s) > 0.33$.

Classification results for each case are visualized in the form of the so-called classification star (see Fig. 5) or in the form of the so-called classification column chart (see Fig. 6).

Aggregation factors have been validated by experiments carried out on a data set with over 1,000 MMPI profiles of women. The quality of classification of cases for described aggregation factors is shown in Table 3. This quality is calculated as a ratio of a number of cases for which a class assigned by a diagnostician is the same as a class indicated by the classification system to a number of all cases.

4 Conclusions

In this chapter, we have described the Copernicus system—for computer-aided diagnosis of mental disorders based on personality inventories. The main attention has been focused on rule-based classification. This part of the tool has been presented more precisely. Our main goal of research is to deliver to diagnosticians and clinicians an integrated tool supporting the comprehensive diagnosis of patients. The Copernicus system is flexible and it can also be diversified into supporting differential diagnosis of profiles of patients examined by means of other professional multidimensional personality inventories.

Acknowledgments This chapter has been partially supported by the grant from the University of Information Technology and Management in Rzeszów, Poland.

References

1. Bazan, J.G., Szczuka, M.S.: The rough set exploration system. In: Peters, J., Skowron, A. (eds.) Transactions on Rough Sets III, Lecture Notes in Computer Science, vol. 3400, pp. 37–56. Springer, Berlin Heidelberg (2005)
2. Butcher, J. (ed.): MMPI: Research Developments and Clinical Application. McGraw-Hill Book Company (1969)
3. Choynowski, M.: Multiphasic Personality Inventory (in Polish). Psychometry Laboratory Polish Academy of Sciences, Warsaw (1964)
4. Cios, K., Pedrycz, W., Swiniarski, R., Kurgan, L.: Data Mining A Knowledge Discovery Approach. Springer, New York (2007)
5. Dahlstrom, W., Welsh, G.: An MMPI A Guide to use in Clinical Practice. University of Minnesota Press, Minneapolis (1965)
6. Dahlstrom, W., Welsh, G., Dahlstrom, L.: An MMPI Handbook, vol. 1–2. University of Minnesota Press, Minneapolis (1986)
7. Duch, W., Kucharski, T., Gomuła, J., Adamczak, R.: Machine learning methods in analysis of psychometric data. Application to Multiphasic Personality Inventory MMPI-WISKAD (in polish). Toruń (1999)
8. Gomuła, J., Paja, W., Pancerz, K., Mroczek, T., Wrzesień, M.: Experiments with hybridization and optimization of the rules knowledge base for classification of MMPI profiles. In: Perner, P. (ed.) Advances on Data Mining: Applications and Theoretical Aspects, LNAI, vol. 6870, pp. 121–133. Springer, Berlin Heidelberg (2011)
9. Gomuła, J., Paja, W., Pancerz, K., Szkoła, J.: A preliminary attempt to rules generation for mental disorders. In: Proceedings of the International Conference on Human System Interaction (HSI 2010). Rzeszów, Poland (2010)
10. Gomuła, J., Paja, W., Pancerz, K., Szkoła, J.: Rule-based analysis of MMPI data using the Copernicus system. In: Hippe, Z., Kulikowski, J., Mroczek, T. (eds.) Human-Computer Systems Interaction. Backgrounds and Applications 2. Part II, Advances in Intelligent and Soft Computing, vol. 99, pp. 191–203. Springer, Berlin Heidelberg (2012)
11. Gomuła, J., Pancerz, K., Szkoła, J.: Computer-aided diagnosis of patients with mental disorders using the Copernicus system. In: Proceedings of the International Conference on Human System Interaction (HSI 2011). Yokohama, Japan (2011)
12. Gomuła, J., Pancerz, K., Szkoła, J., et al.: Classification of MMPI profiles of patients with mental disorders—Experiments with attribute reduction and extension. In: Yu, J. (ed.) Rough

Set and Knowledge Technology, Lecture Notes in Artificial Intelligence, vol. 6401, pp. 411–418. Springer, Berlin Heidelberg (2010)

13. Gomuła, J., Pancerz, K., Szkoła, J.: Rule-based classification of MMPI data of patients with mental isorders: experiments with basic and extended profiles. Int. J. Comput. Intell. Syst. **4**(5) (2011)

14. Greenes, R.: Clinical Decision Support: The Road Ahead. Elsevier, Amsterdam (2007)

15. Lachar, D.: The MMPI: Clinical Assessment and Automated Interpretations. Western Psychological Services, Fate Angeles (1974)

16. Płuzek, Z.: Value of the WISKAD-MMPI test for nosological differential diagnosis (in polish). The Catholic University of Lublin (1971)

17. Witten, I.H., Frank, E.: Data Mining: Practical Machine Learning Tools and Techniques. Morgan Kaufmann (2005)

An Adaptive Approach for Noise Reduction in Sequences of CT Images

Veska Georgieva, Roumen Kountchev and Ivo Draganov

Abstract CT presents images of cross-sectional slices of the body. The quality of CT images varies depending on penetrating X-rays in a different anatomically structures. Noise in CT is a multi-source problem and arises from the fundamentally statistical nature of photon production. This chapter presents an adaptive approach for noise reduction in sequences of CT images, based on the Wavelet Packet Decomposition and adaptive threshold of wavelet coefficients in the high frequency sub-bands of the shrinkage decomposition. Implementation results are given to demonstrate the visual quality and to analyze some objective estimation parameters such as PSNR, SNR, NRR, and Effectiveness of filtration in the perspective of clinical diagnosis.

Keywords CT image · Noise reduction · Wavelet transformations · Adaptive threshold

1 Introduction

In Computed Tomography, the projections acquired at the detector are corrupted by quantum noise. This noise propagates through the reconstruction to the final volume slices. This noise is not independent of the signal. It's Poisson distributed

V. Georgieva (✉) · R. Kountchev · I. Draganov
Department of Radio Communications and Video Technologies, Technical University of Sofia, Boul. Kl. Ohridsky 8, 1797 Sofia, Bulgaria
e-mail: vesg@tu-sofia.bg

R. Kountchev
e-mail: rkountch@tu-sofia.bg

I. Draganov
e-mail: idraganov@tu-sofia.bg

B. Iantovics and R. Kountchev (eds.), *Advanced Intelligent Computational Technologies and Decision Support Systems*, Studies in Computational Intelligence 486, DOI: 10.1007/978-3-319-00467-9_4, © Springer International Publishing Switzerland 2014

and independent of the measurement noise [1]. We cannot assume that, in a given pixel for 2 consecutive but independent observation intervals of length T, the same number of photons will be counted. The measurement noise is additive Gaussian noise and usually negligible relative to the quantum noise. It comes from the motion of patient [1].

The methods of noise reduction can be categorized as spatial domain methods and transform domain methods. In the case of spatial domain methods, the noise in original image is reduced within the spatial domain. It employs methods like neighborhood average method, Wiener filter, center value filter and so on. In the transformations field the image is transformed, and the coefficients are processed to eliminate noise [2, 3]. This methods generally have a dilemma, between the rate of noise reduction and holding of image edge and detail information. Recent wavelet thresholding based denoising methods proved promising, since they are capable of suppressing noise while maintaining the high frequency signal details. The wavelet thresholding scheme [4], which recognizes that by performing a wavelet transform of a noisy image, random noise will be represented principally as small coefficients in the high frequency sub-bands. So by setting these small coefficients to zero, will be eliminated much of the noise in the image.

In this chapter, we purpose to reduce the noise components in the CT images on the base of 2D wavelet packet transformations. To improve the diagnostic quality of the medical objects some parameters of the wavelet transforms are optimized such as: determination of best shrinkage decomposition, threshold of the wavelet coefficients and value of the penalized parameter of the threshold. This can be made adaptively for which image in the sequence on the base of calculation and estimation of some objective parameters.

The chapter is arranged as follows: In Sect. 2 is given the basics stages of the proposed approach; in Sect. 3 are presented some experimental results, obtained by computer simulation and their interpretation and Sect. 4—the Conclusion.

2 Basic Stages for Noise Reduction in CT Image

The general algorithm for noise reduction based on wavelet packet transform contains the following basic stages:

- Decomposition of the CT image;
- Determination of the threshold and thresholding of detail coefficients;
- Restoration of the image;
- Estimation of filtration.

2.1 Decomposition of the Image

The scheme for obtaining the Wavelet packet decomposition (WPD) for noise reduction is shown on Fig. 1.

The wavelet packet analysis is a generalization of wavelet decomposition that offers a richer image analysis. The wavelet packet methods for noise reduction give a richer presentation of the image, based on functions with wavelet forms, which consist of 3 parameters: position, scale and frequency of the fluctuations around a given position. They propose numerous decompositions of the image, that allows estimate the noise reduction of different levels of its decomposition. For the given orthogonal wavelet functions exists library of bases, called wavelet packet bases. Each of these bases offers a particular way of coding images, preserving global energy, and reconstructing exact features. Based on the organization of the wavelet packet library, it is determinate the decomposition issued from a given orthogonal wavelets. A signal of length $N = 2^L$ can be expand in α different ways, where α is the number of binary sub trees of a complete binary tree of a depth L. The result is $\alpha \geq 2^{N/2}$ [4]. As this number may be very large, it is interesting to find an optimal decomposition with respect to a conventional criterion. The classical entropy-based criterion is a common concept. It is looking for a minimum of the criterion. In case of denoising the 2D joint entropy of the wavelet co-occurrence matrix is used as the cost function to determine the optimal threshold. In this case 2D Discrete Wavelet Transform (DWT) is used to compose the noisy image into wavelet coefficients [5].

We propose in the chapter another adaptive approach. The criterion is a minimum of three different entropy criteria: the energy of the transformed in wavelet domain image, Shannon entropy and the logarithm of energy [6].

Looking for best adaptive shrinkage decomposition to noise reduction, two important conditions must be realized together [7]. The conditions (1) and (2) are following:

$$E_k = \min, \quad \text{for } k = 1, 2, 3, \ldots, n \tag{1}$$

where E_k is the entropy in the level **K** for the best tree decomposition of image.

$$s_{ij} \geq T \tag{2}$$

where s_{ij} are the wavelet coefficients of image **S** in an orthonormal basis, **T** is the threshold of the coefficients.

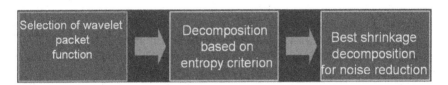

Fig. 1 Scheme for obtaining the WPD for noise reduction

To select the level of the shrinkage decomposition is the next step in the algorithm.

2.2 Determination of the Threshold and Thresholding of Detail Coefficients

By determination of the global threshold it is used the strategy of Birge-Massart [8]. This strategy is flexible. It uses spatial-adapted threshold, which allows to determinate the thresholds in three directions: horizontal, vertical and diagonally. The wavelet thresholding procedure removes noise by thresholding only the wavelet coefficients of the detail sub-bands, while keeping the low resolution coefficients.

The threshold can be hard or soft. The soft-thresholding function takes the argument and shrinks it toward zero by the threshold. The soft-thresholding method is chosen over hard-thresholding, because it yields more visually pleasant images over hard-thresholding. To become more precisely determination of the threshold for noise reduction in the image we can penalize adaptively the sparsity parameter α [9]. Choosing the threshold too high may lead to visible loss of image structures, but if the threshold is too low the effect of noise reduction may be insufficient.

All these elements of the procedure for noise reduction can be determined on the base of the calculated estimation parameters. PSNR and E_{FF} values are higher for better denoised CT image where the value of NRR is lower.

2.3 Restoration of the Image

The restoration of the image is on the base on 2D Inverse Wavelet Packet Transform. The reconstructions level of the denoised CT images in the sequence is different for each image and is dependent on the level of its best shrinkage decomposition.

2.4 Estimation of Filtration

In this stage are analyzed some objective estimation parameters such as Peak signal to noise ratio (PSNR), Signal to noise ratio in the noised image (SNR_Y), Signal to noise ratio in the filtered image (SNR_F), Noise reduction rate (NRR), and Effectiveness of filtration (E_{FF}), and SSIM coefficient in the perspective of clinical diagnosis. All adaptive procedures in the proposed algorithm are made automatically, based on calculated estimation parameters. PSNR and E_{FF} values are higher for better denoised CT image where the value of NRR is lower.

3 Experimental Results

The formulated stages of processing are realized by computer simulation in MATLAB environment by using WAVELET TOOLBOX. The image data consists of grayscale CT-slices images of the liver of size 512×512 pixels, 16 bits in DICOM format. They are 2,367 images, archived in 12 series. In analyses is presented a sequence of 6 CT images.

The simulation is made with additive Poisson noise with intensities values between 0 and 1 and corresponding to the number of photons divided by 10^{12}.

All computations were performed using Coiflet wavelet packet function. The best results for the investigated sequence of 6 CT images are obtained by the third, fourth and fifth level of the shrinkage decompositions. They are obtained by using the Shannon entropy criteria. By using of the log energy and energy criteria the effectiveness of the filtration is smaller. For noise reduction is used soft threshold.

In Fig. 2 are shown the best shrinkage decompositions for CT image IM00004 (on level 4), obtained by using the three types of entropy criterion.

In the chapter are analyzed some quantitative estimation parameters: PSNR, Signal to noise ratio in the noised image (SNR_Y), Signal to noise ratio in the filtered image (SNR_F), Effectiveness of filtration (E_{FF}) Noise reduction ratio (NRR) and SSIM.

Table 1 contains the simulation results from the noise reduced CT image IM0004 (on level 4), obtained by using the three types of entropy criterion.

By using of the log energy and energy criteria the values of PSNR, SNR_F and E_{FF} are lower and the effectiveness of the filtration is smaller.

Table 2 contains the obtained results from the noise reduced sequence of the 6 CT images.

In order to quantify how much noise is suppressed by the proposed noise reduction approach, the noise reduction rate is computed. The obtained average results for NRR are around 0.3 and shows that the noise is three times reduced. The calculated values of SSIM for all images are around 0.7.

The simulation results show that by increasing the level of the best shrinkage decomposition the values of SNR_F and E_{FF} increase too, but the value of PSNR decreases.

The obtained result can be compared with other adaptive methods for noise reduction in CT images of the liver, based on wavelet discrete transform (DWT) [10]. In this case the NRR plot shows that noise is reduced by approximately 0.5 or around two times. Such comparison is very difficult through large variations of liver geometry between patients, the limited contrast between the liver and the surrounding organs and different amplitude of noise. Other sides the properties of

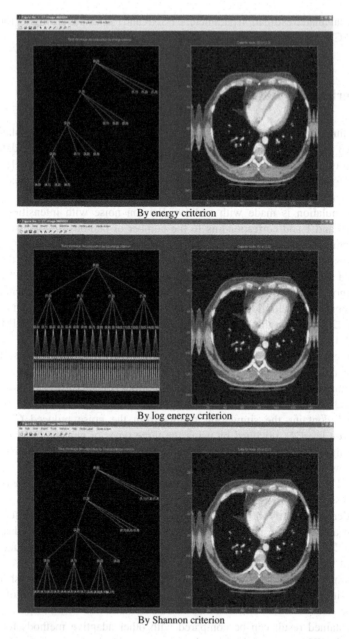

By energy criterion

By log energy criterion

By Shannon criterion

Fig. 2 The best shrinkage decompositions for CT image IM00004 on level 4 by three different type of entropy criterion. **a** By energy criterion. **b** By log energy criterion. **c** By Shannon criterion

the images are different, such as resolution and depth. In many case the estimation was made only on the base of particular parameters such as PSNR or SNR [11, 12].

Table 1 Simulation results for denoised CT image IM00004 by using three types of entropy criterion

Entropy criterion	PSNR (dB)	SNR_Y (dB)	SNR_F (dB)	E_{FF} (dB)
Energy	28.5085	16.7527	18.1347	1.3820
Log energy	28.3474	16.7527	18.0683	1.3156
Shannon	28.5864	16.7527	18.1647	1.4120

Table 2 Simulation results for denoised sequence of the 6 CT images

Image	Level	PSNR (dB)	SNR_Y (dB)	SNR_F (dB)	E_{FF} (dB)
IM00001	3	28.8910	17.2103	18.3181	1.1076
IM00003	4	28.6506	16.8878	18.6506	1.3561
IM00004	4	28.5864	16.7527	18.1647	1.4120
IM00005	5	28.3706	16.5655	18.0115	1.4446
IM00008	5	28.2527	16.0946	17.8386	1.7440
IM00010	5	27.9229	16.1185	18.0895	1.9710

A visual presentation of the obtained results for the processed sequence of the 6 CT images can be seen in the next figures below. They all present: the original CT image, the filtered image and the energy of filtered noise (Figs. 3, 4, 5, 6, 7 and 8).

The implemented studying with sequence of real CT images and the obtained experimental results have shown that:

- The proposed effective approach for noise reduction based on WPD can be adaptive applied for every CT image-slice in the sequence;
- It's necessary to do this in regard to different rate of noise in the images, which varies depending on penetrating X-rays in a different depth of the anatomical structures;
- The parameters of this filter can be automatically and separately for every CT image in the sequence determinate on the base of calculated objective estimations;
- A complementary adjustment can be made in the case of the level of shrinkage decomposition and the sparsity parameter α of the penalized threshold. It can be

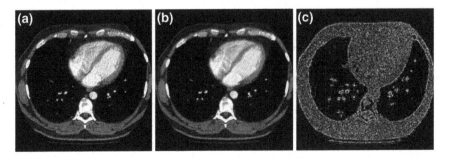

Fig. 3 CT image IM00001: **a** original image, **b** filtered image, and **c** energy of the reduced noise

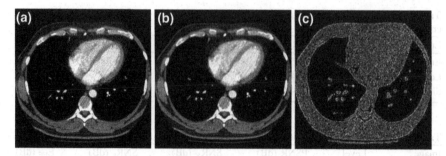

Fig. 4 CT image IM00003: **a** original image, **b** filtered image, and **c** energy of the reduced noise

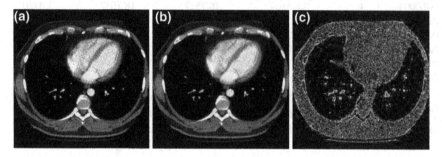

Fig. 5 CT image IM00004: **a** original image, **b** filtered image, and **c** energy of the reduced noise

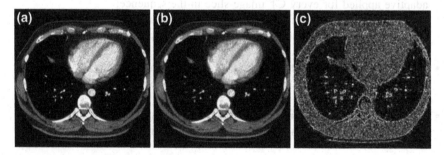

Fig. 6 CT image IM00005: **a** original image, **b** filtered image, and **c** energy of the reduced noise

applied for selected ROI of the CT images too, but depending on the specific character of the region;

• The noise reduction process can't be applied by differences in the inter sequence images in the case of reiteration of elements in the same position for loss of specific medical information.

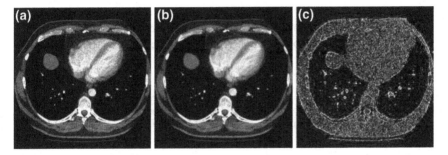

Fig. 7 CT image IM00008: **a** original image, **b** filtered image, and **c** energy of the reduced noise

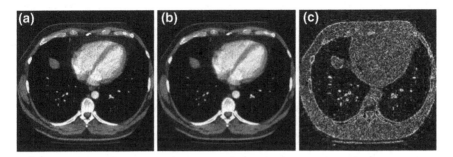

Fig. 8 CT image IM00010: **a** original image, **b** filtered image, and **c** energy of the reduced noise

4 Conclusions

In the chapter is proposed a new and effective adaptive approach for noise reduction in sequences of CT-slice images. It's based on wavelet packet transformation. The implemented algorithm provides a basis for further investigations in several directions:

- For elimination of more noised CT-slice images from the investigated sequence which are in near neighborhoods (have the same diagnostic information) of images with lower rate of noise. So can be decreased the volume of image data in the sequences while only the better images will be kept in the database;
- The denoised sequence of CT images can be used for better visualization in 3D reconstruction;
- The denoised sequence of CT images can be post processed with methods for segmentation, based on adaptive WPD for detection of specifically regions and edges with more diagnostic information;
- Some statistical characteristics such as histogram of the images and the functions of noise distribution can be analyzed for full automatically reduction of the noise;

- The denoised sequence of CT images can be effective compressed for decreasing the image database for further transmission and archiving;
- The obtained image database can be easily implicated for classification of different diseases.

Acknowledgments This chapter was supported by the Joint Research Project Bulgaria-Romania (2010–2012): "Electronic Health Records for the Next Generation Medical Decision Support in Romanian and Bulgarian National Healthcare Systems", DNTS 02/19.

References

1. Smith, M., Docef, A.: Transforms in telemedicine applications. Kluwer Academic Publishers, Boston (1999)
2. Athhanasiadis, T., Wallace, M., Kapouzis, K., Kollias, S.: Utilization of evidence theory in the detection of salient regions in successive CT images. Oncol Rep **15**, 1071–1076 (2006)
3. Donoho, D., Johnston, I.: Ideal spatial adaptation via wavelet shrinkage. Biometrika **81**, 425–455 (1994)
4. Donoho, D., Johnston, I.: Adapting to unknown smoothness via wavelet shrinkage. Am. Stat. Assoc. **90**, 1200–1224 (1995)
5. Zeyong, S., Aviyente, S.: Image denoising based on the wavelet co-occurance matrix. IEEE Trans. Image Proc. **9**(9), 1522–1531 (2000)
6. Coifmann, R., Wickerhauser, M.: Entropy based algorithms for best basis selection. IEEE Trans. Inf. Theory **38**, 713–718 (1992)
7. Georgieva, V., Kountchev, R.: An influence of the wavelet packet decomposition on noise reduction in ultrasound images. In: Proceedings of International Scientific Conference on Information, Communication and Energy systems and Technology, pp. 185–188, Sofia, Bulgaria (2006)
8. MATLAB User's Guide. Accessed at:www.mathwork.com
9. Georgieva, V.M.: An Approach for computed tomography images enhancement. Electron. Electr. Eng. **2**(98), 71–74 (2010) (Kaunas: Technologija)
10. D. Bartuschat, A. Borsdorf, H.Koestler, R. Rubinstein, M. Stueurmer. A parallel K-SVD implementation for CT image denoising. Fridrich-Alexander University, Erlangen (2009)
11. Hyder Ali, S., Sukanesh, R.: An improved denoising algorithm using curvlet thresholding technique for medical images. Int. J. Adv. Comput. (IJAC) **2**(2), 83–91 (2010)
12. Amijad Ali, S., Srinivasan, S., Lalkishore, K.: CT Image denoisung technique using ga aided window-based multiwavelet transformation and thresholging with incorporation of an effective quality enhancement method. Int. J. Digit. Content Technol. Appl. **4**, 73–87 (2010)

Intelligent Predictive Diagnosis on Given Practice Data Base: Background and Technique

George Isoc, Tudor Isoc and Dorin Isoc

Abstract Medical diagnosis is a model of technical diagnosis for historical reasons. At this point, technical diagnosis results as a set of information processing techniques to identify technical faults can be extent to the medical field. Such situation is referring to the modeling of the cases and to the use of the information regarding the states, medical interventions and their effects. Diagnosis is possible not only through a rigorous modeling but also through an intelligent use of the practice bases that already exist. We examine the principles of such diagnosis and the specific means of implementation for the medical field using artificial intelligence techniques usual in engineering.

1 Introduction

Artificial Intelligence is an area that affords to suggest general solutions to be universally used. One such area is that of the diagnosis to be applied to technical systems but also to humans. Despite the fact that the medicine was the starting point, the use of simplified models has enabled the important development of technical systems.

This is proven by the development of reasoning and modeling techniques as parts of intelligent systems [1–11]. These are more diverse, more abstract, but with

G. Isoc (✉)
Clinical Emergency Hospital Oradea, Oradea, Romania
e-mail: George.Isoc@yahoo.com

T. Isoc
Technical University of Cluj-Napoca, Cluj-Napoca, Romania
e-mail: Tudor.Isoc@eps.utcluj.ro

D. Isoc
Integrator Consulting Ltd, Cluj-Napoca, Cluj-Napoca, Romania
e-mail: Dorin.Isoc@yahoo.com

B. Iantovics and R. Kountchev (eds.), *Advanced Intelligent Computational Technologies and Decision Support Systems*, Studies in Computational Intelligence 486, DOI: 10.1007/978-3-319-00467-9_5, © Springer International Publishing Switzerland 2014

increasing restrictions on the application. In principle, the restrictions of the application refer to the need for developed intelligent systems to be able to learn and further for their bases of knowledge to be validated. This tendency makes artificial systems, as decision-making systems including models and algorithms of Artificial Intelligence, factitious and, the associated theory, difficult to apply.

There are rare researches that leave the field of theory and approach the essence of the methodology of human diagnosis [12–14] in order to make it primarily a low-bias activity. The achieved intelligent systems [15, 16] are usually individual applications with a small area of extension and they are not well received in many medical environments.

Problems on the diagnosis remain at the starting point, weather it is a technical or human one. It is more important to set a reasoning technique than to do a complete description of the viable information that can be used for making a diagnosis. Finally the diagnosis should be put into a dynamics of its own statements of functional abnormality. The vision change introduces predictive capacity resulting from the diagnosis, prediction or forecast which the human factor will never completely ignore.

The chapter aims to set the principles of intelligent prediction diagnosis of technical systems for intelligent automatic medical diagnosis, the development of the database and of the exploitation algorithm necessary to assist the medical decision.

In the first chapter, we introduce the principles of automatic diagnosis of the technical systems and we set the limitations that occur when the object of the diagnosis is the human patient. Using these principles we suggest a representation model of the practice base and further, an algorithm of exploitation of practice base for an assisted decision. A series of interpretations are made and further we draw concluding remarks about the contribution of the suggested technique.

2 Is a Fault State or A Disease an Abnormal Condition?

Although was born at the same time as the human being, the disease status is also reflected in specific forms in the artificial world of machines. The similarities between the two states have made that the language used in the technical field to be used in the medical field. And here it seemed that the similarities were exhausted.

Cornerstone of medical diagnosis (medical diagnostic) is the collecting of interest information through discussion or dialog with the patient. Further, it come the tests and examinations based on measurements and equipments, followed once again by discussion with the patient, and then the prescription of treatment followed by the evaluation which again is followed by a discussion with the patient.

In technical world, one essential element is missing: the discussion with the patient! There for the "technical" doctor does not talk to anyone and has no place from which to collect any specific information and can not confirm assumptions and effects of the medication. Another important detail: The *technical patient* does not come to the *technical doctor* to complain about anything!

Despite these essential differences, one can find many similarities between the two situations.

First, it is necessary to set the terms. For the patient, we talk about *disease* or *condition* while for the technical system, we talk about *fault*. In both cases, the symptom is the clue indicating the presence of an abnormal state, is it disease or fault.

Further, we will indicate the elements that allow a definition of the fault state [17] and the connections with those features that make the patient a form/a part of a kind of *system*.

From the medical point of view, beyond any ethical limitations, the disease is a kind of problem in the general sense of the information. This is to say that the disease appears as an incomplete collection of information. The gap, discovered with the help of the other information we acquire in the collection, is the informational solution to the problem.

The collection of information is made up of what the doctor knows as a result of the studies and professional investigations, what the doctor knows as a result of personal readings, what the doctor knows from accumulation and filtering of daily experience. The collection is filled with what the doctor may acquire through the use of investigative techniques and technical means such as appliances and means of investigation.

Finally, the collection is completed with information that the doctor acquires, either directed or not, following the dialog with the patient.

If we now make an analysis of the disease as a fault condition for the normal function of the human being, we can identify issues that can be interpreted.

For a fair comparison, it is necessary that the technical system is seen as an ordered collection of objects made in order to achieve a purpose. Another common element regarding the information is that it says that all key parameters of the system and of the patient supposed to have an evolution over time. We thus acknowledge that the technical system model and the patient model are dynamic.

The whole key parameter values group will be known as the state of the dynamic system. Key parameters as functions of time will be known as variables of the dynamic system.

We notice that by doing this, the technical system and the patient are taken together with their dynamics and disease, fault or defective condition are part of these dynamics.

In order to work with the concept of fault state it is necessary to define it as detailed is possible. This definition will be given as a specification, i.e. defining the state in a descriptive way as using the Rule 1 to Rule 9.

Rule 1 *The fault state is a possible state of the system.*

Technical description: The defined state of a system represents the number of sets of values of its variables for all the time moments.

The fault state is characterized through the same variables. The fault state appears in some time moments.

Medical environment: Identical as in the technical system.

Rule 2 *The fault state is an abnormal state.*

Technical description: Each system has a certain finality established. Functionally speaking, this means that from the number of possible states for the system, one part is represented by the number of states which obey imposed finalities. We'll say that these states are called normal states or, more correctly, normal functional states. We'll name all the other possible states abnormal states as oppose to the normality. There is however one part of the abnormal state that is not only representing the finality of the system but it represents the contrary of the normal state or even a totally user inconvenient. Among these possible states we also find *the fault state*. The normality of a state is defined by the human factor. Normality is a problem of convention and experience.

Medical environment: As a human being, the patient can not be assigned with a purpose. The other details still stand. The normality of a state is learned or is defined by a medical standard.

Rule 3 *The fault state is a stable state.*

Technical description: The variables which define the fault state reveal the snapshot states of a system. The process of revealing a state at a certain moment can be assimilated to a measurement process. The conditions imposed to the measurement assume the stability of the state that is the limited character of the value, and also of the variation of the value at a given moment in time. It follows that *the fault state is a stable state.*

Medical environment: It must be taking in consideration that we are dealing with a living organism. It is necessary to separate the fault state, ie sickness form accident. Although they appear almost the same, the two situations are different considering that the accident doesn't have stability. It assumes that the intervention to remove the state of disease is promptly. If the state of disease is not removed in time it can end up to irreversible changes. Example: A child who is feverishly will dehydrate and soon reach the critical state.

Rule 4 *The fault state has a certain time horizon.*

Technical description: The fault state is a possible state of the machine. It's also a part of the "history" of the machine. The fault state also has a beginning and an ending. In this way, *the fault state has a certain time horizon.* In addition, every fault state has a certain dynamic. As the fault state is stable, it follows that the duration of the fault state is longer than the systems transitional condition. It follows that we have two types of faults. The first type is that of the faults with immediate effect. This type of fault can fall into the category of accidents. Their mere appearance means the exhaustion of the fault state. The second type is of *the faults with continuous effect.* Unlike the previous case, in these cases we find that the faults' effects are present for a significant time period.

Medical environment: Any disease is known through it's starting event or symptoms, through the signs of evolution and a normal duration of the disease and then through the signs that can predict the healing process. We notice that most of the diseases with acute manifestation can be included in this category.

Rule 5 *The fault state may be due to a functional or structural cause.*

Technical description: By the causes that are determining it, the fault state may be a consequence of an internal disturbance such as the modification of a physical parameter. The effect manifests on the system state which becomes the fault state. The fault state has the information on the way the system works but also on the causes that produced it.

Medical environment: Without exceptions the rule is valid.

Rule 6 *The fault state can be foreseen almost anytime.*

Technical description: Any system exists in the boundaries of a human experience. This experience involves the accumulation of experience by default, by sequencing analysis and troubleshooting. Especially those symptoms causing functional faults, without leading to destruction prove to be of large interest. Functional faults determine behaviors outside the technical system performance requirements.

Medical environment: For the patient the functional performances are reduced to normal parameter values and to the existence of the state of comfort and wellbeing. It is difficult to precisely define these conditions.

Rule 7 *The fault state can not always be built out of measurable information.*

Technical description: In real terms, defining the set of state variables must coincide, in theory even entirely, with the set of measurable variables of the physical system. While not all determinant variables of the physical system can be measured, some information that could define the state will not be accessible to an automatic diagnosis. It is there for required that some *state variables of the machine*, thus of the fault state *to be constructed* (estimated) or *obtained by other means*, such as, for example, information from the human factor.

Medical environment: The situation is specific to living matter where mea- surements can not be achieved either because of lack of access or lack of sensitive equipment

Rule 8 *The fault state is always outside the possibilities of the control system.*

Technical description: A technical system is difficult to be conceived without human or automatic control systems. Their role is to keep certain variables in the user's desired limits. The control system will act on some of the abnormal states of the machine in order to compel them to determine the output values within the limits imposed purpose. It results that the control system can act on some state variables that tend to become abnormal in the sense of forcing them to become normal. If we consider these states as fault states is wrong because their infor- mation is already modified by the intervention of the control system. Thus the diagnostic system will treat as fault states only those abnormal states which are outside the control systems performances.

Medical environment: It is proven that worldwide we can be find a lot of cyber connections, including the ones for control or for self-control.

Rule 9 *The fault state does not always affect the controlled systems performances.*

Technical description: A reciprocal like: a fault state does not necessarily affect the performance of the controlled system is not always true. The reason for this is that an inspection will mask the existence of a fault by constraints imposed on a fault state considered only as an abnormal state: the system of control can not be advised on the reasons of the abnormal state, ie an external or internal one.

Medical environment: This situation is described by the fact that the general bad state is not the same for all patients.

Restriction 1 *In a technical system, once a fault occurred it does not self-repair.*

Medical environment: Unlike technical systems, living organisms have self-control mechanisms that can save them from different situations without medical interventions.

Restriction 2 *In a technical system, an occurred fault can maintain itself or develop into other faults with more important intensities.*

Medical environment: Despite the fact that in the living world self-control exists, the disease can degenerate in complications, similar to significant faults.

Restriction 3 *A diagnosis system is intended to identify the fault state in relation to a normal state defined or confirmed as such by the user.*

Medical environment: Normality or health status is learned by each person. The disease can occur only in situations where the normal state is abandoned or is not found.

Restriction 4 *A diagnosis system can be used for identification, isolation, prevention or prediction of the fault state.*

Medical environment: The above actions are the very actions that a good health system is developing.

Restriction 5 *The upper limits of the diagnosis system performance are imposed by the lower limits of the control system performance of the variables included in the state of the machine.*

Medical environment: Illness of a patient with good general condition is always detected only in a serious stage because his body is able to fight or compensate the early stages of the disease.

Restriction 6 *The diagnosis system can not act before the control system proved to be inefficient in its action to constrain the variables ordered even if for proving this situation a priori threshold is required.*

Medical environment: A condition can be detected only after the immune system has been exceeded. We can deduce from this the importance of preventive actions that may give clues that can be interpreted before signs of disease.

Restriction 7 *For systems with human operators monitoring, the diagnosis system performance limit is given by the limit of power expressed as the human operators' experience.*

Medical environment: The medical system requires a power expressed by medical staff qualifications, through its experience and the technology available used for diagnosis. Monitoring is conducted here only with human operators.

Restriction 8 *The diagnosis system is always located at a higher level of control, at the level of supervision.*

Medical environment: There is a medical system that has a specific function within the organization of society and its place is established independently of the other functions of society.

Restriction 9 *Each fault state shall be provided with its specific time horizon.*

Medical environment: Every disease has a period of development and of recovery established by the medical practice over time.

Restriction 10 *The fault state can not be defined only outside the supervised system.*

Medical environment: No medical staff recommends self-medication and self diagnosis.

Restriction 11 *A state which a technical system did not experience is a potential question mark over its future. However, it is difficult to claim that a machine, who oversees the system, is able to know all possible states of a given technical system.*

Medical environment: The restriction is valid only on abnormal states whereas human beings have an organic development process. It is natural that the man throughout his life meets new states that are not necessarily abnormal or harmful. For limited periods of time, the above restriction is valid and may be a reason for the action of the diagnosis system.

Restriction 12 *The decision on whether a state is normal or abnormal is taken in a context where the human operator intervenes or provides new information to filter common information.*

Medical environment: The decision on a state of disease is always confirmed by a human specialist.

In the detailed descriptive definition some restrictions are required that affect the use of information on the fault state and also the appropriate state. The main restrictions are presented and briefly argued from Restriction 1 to Restriction 12.

In conclusion, for the design and development of predictive technical diagnostic systems we can enunciate a fundamental proposition:

In a technical system, a functional fault always occurs at random but then always has a deterministic and predictable evolution.

Now it is necessary to introduce the major technical diagnosis restriction in its predictive version:

The predictive diagnosis role is to identify the fault state before producing important technical and economic effects.

For the technical systems the predictive diagnosis is important. It has anticipatory character and it is able to keep the technical systems functionality within acceptable limits.

The similarity with the medical diagnosis is immediate even if the technical and economic effects need to be replaced by "impairment of health".

3 Diagnosis on a Practice Base

Next the introduced and discussed concepts will be used in order to extend predictive diagnostic techniques in the medical field. First we'll make a description of the practice (experience) base.

3.1 Practice Base and Its Description

Any practice that refers to the same kind of facts can be uniformly described by a set of information. If we take into account the fact that the disease is a "fault state" meaning that it respects the set rules from Rule 1 to Rule 9, then the facts can be described:

- Through *context* which is a limited but sufficiently large collection of historical data that are not determined by updated information or updates.
- Through *situation* which is a minimum stock of knowledge that can define the current status of the patient for the doctor through accepted parameters and values. It is noted that a statement always refers to a limited time horizon and can be normal or critical, but this is defined in relation to the physician's opinion and the patient statements.

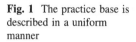
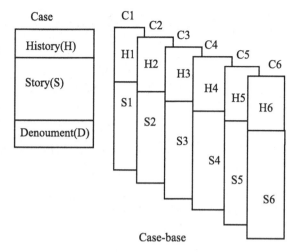

Fig. 1 The practice base is described in a uniform manner

Typically the practice base as in Fig. 1 has cases (*C*) and each case is characterized by a history (*H*), a story (*S*) daily updated, including the last situation and an evolution (*D*) which is a desired future state or a state that the patient can reach as effect of a predicted evolution of his disease (Fig. 2).

More than that, the *S* story can be decomposed in episodes. Each episode is a disease or fault case as in Fig. 3.

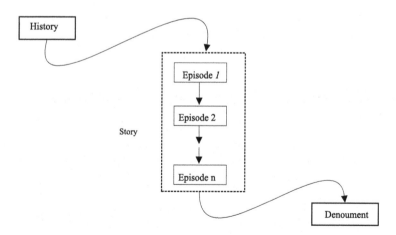

Fig. 2 Temporary evolution (*S*) of each case appears as a chain of episodes while the global, static description (*H*) global of the case system or patient) is unique. Each case has a final outcome called *denouement* (*DD*) at some point of time. This time is not necessarily the end of the case as existence but it is the end of known existence, usually identical to an earlier time or the present time moment

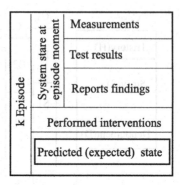

Fig. 3 The story of each case (system) consists of episodes where the doctor intervened to remedy health after diagnosis. Each episode has a certain duration, typical to the patient (system). The episode is identified by an initial state, a set of interventions and as a result, a new state where the effects of the fault or disease are eliminated or diminished. This new state is expected as anticipated normality state by the diagnostic system (doctor) by prediction

3.2 Diagnosis Intelligent Algorithm

On the fault state description as in rules set of Rule 1 to Rule 9 and the restrictions of the set of Restriction 1 to Restriction 12 it is possible to build a diagnosis predictive algorithm based on the principle that states in order to get similar effects (positive evolution) are retrieved with the closest case in a given practice base and apply similar solutions.

The algorithm given in Table 1 offers the doctor the support based on the similarity between a current case and a large set of previous practices.

Table 1 The story of each case assumes episodes where the doctor acted to improve the health state after a diagnosis

Step 1:	Fill in the current case with the last event of the story with measurements, results, tests and report findings
Step 2:	do $i = 1$ and $k = 0$
Step 3:	*WHILE* there still are cases on the experience basis *REPEAT*
Step 3.1:	*IF* the case exemplar looks like the case experience of i *THEN*
Step 3.1.1:	Remember case i
Step 3.1.2:	Do $k = k + 1$
Step.3.2:	Do $CA_k = C_i$
Step 3.3:	*END IF*
Step 3.4:	DO $i = i + 1$
Step 4:	*END REPEAT*
Step 5	Provide list of similar cases in support of alternative medical treatments

4 Interpretations

We find that the practice base should not belong to only one doctor but it is necessary that the description of the cases to be uniform. Without going into details, the *H* history contains information defining the individual in terms of family, employment, living conditions. The health state has an continuous in time evolution, the story (*S*), situated among disease states where the intervention of medical staff is necessary. The practice proves that individuals with similar history (*H*) and story (*S*) have similar states of evolution. For young individuals, the states already occurred only by similar previous cases become predicted disease or health states.

This is the way to use the practice as a manner to predict future states. So the diagnosis becomes *predictive diagnosis*. The nature of information characterizing the complex systems and also the patient gives to diagnosis the features of *intelligent system*.

We also find that the question of similarity between cases is present, but it may only be a simple fixed threshold value when implementing the technical solution.

5 Concluding Remarks

The chapter is intended to introduce the principles of intelligent predictive diagnosis and to describe a technique of assisted medical diagnosis but based on the technical systems and artificial intelligence principles.

After a description of the fault state through a series of restrictions and rules underlying the construction of an automatic diagnosis system, the practice base is being shaped followed by its exploitation algorithm. We insists on the fact that from the standpoint of artificial intelligence and those presented here, the important part is the similarity between the examined case and the set of medical experience in the experience base. It is also very important that the whole diagnosis system is one that assists the doctor and not one that replaces the doctor.

The contribution of this work particularly is to define the methodology of cases representation. This representation will allow the use of the knowledge contained in similar cases processed by intelligent predictive diagnosis systems in order to predict health or disease state.

References

1. Ercal, F., Chawla, A., Stoecker, W.V., Lee, H.C., Moss, R.H.: Neural network diagnosis of malignant melanoma from color images. IEEE Trans. Biomed. Eng. **41**(9), 837–845 (1994)
2. van Ginneken, B., ter Haar Romeny, B.M., Viergever, M.A.: Computer-aided diagnosis in chest radiography: A survey. IEEE Trans. Med. Imag. **20**(12), 1228–1241 (2001)

3. Tan, K.C., Yu, Q., Heng, C.M., Lee, T.H.: Evolutionary computing for knowledge discovery in medical diagnosis. Int. Artif. Intell. Med. **27**, 129–154 (2003)
4. Kononenko, I.: Inductive and bayesian learning in medical diagnosis. Int. J. Appl. Artif. Intell. **7**(4), 317–337 (1993)
5. Carpenter, G.A., Markuzon, N.: ARTMAP-IC and medical diagnosis: Instance counting and inconsistent cases. In: Technical Report CAS/CNS-96-017, Boston University Center for Adaptive Systems and Department of Congnitive and Neural Systems (1996)
6. Szolovits, P., Pauker, S.G.: Categorial and probabilistic in medical reasoning in medical diagnosis. Artif. Intell. **11**, 115–144 (1978)
7. Aamodt, A., Plaza, E.: Case-based reasoning: foundational issues, methodological variations, and system approaches. In: Aamodt, A., Plaza, E. (1994); Case-Based Reasoning: Foundational Issues, Methodological Variations, and System Approaches. AI Communications. IOS Press, vol. 7(1), pp. 39–59 (1978)
8. Quinlan, J.R.: Induction of decision trees. Mach. Learn. **1**, 81–106 (1986)
9. Zhou, Z.H., Jiang, Y.: Medical diagnosis with C4.5 rule preceded by artificial neural network ensemble. IEEE Trans. Inf. Technol. Biomed. **1**(7), 37–42 (2003)
10. Peña-Reyes, C.A., Sipper, M.: A fuzzy-genetic approach to breast cancer diagnosis. Artif. Intell. Med. **17**, 131–155 (1999)
11. Kononenko, I.: Machine learning for medical diagnosis: History, state of the art and perspective. Artif. Intell. Med. **23**(1), 89–109 (2001)
12. Sahrmann, S.A.: Diagnosis by the physical therapist: A prerequisite for treatment—A special communication. Physical Therapy **68**(11), 1703–1706 (1988)
13. Peng, Y., Reggia, J.A.: Plausibility of diagnostic hypotheses: The nature of simplicity, In: AAAI-86 Procedings (1986).
14. Croskerry, P.: The importance of cognitive errors in diagnosis and strategies to minimize them. Acad. Med. **78**(8), 775–780 (2003)
15. Clancey, W.J., Shortliffe, E.H., Buchanan, B.G.: Intelligent computer-aided instruction for medical diagnosis, In. Proc Annu Symp Comput Appl Med Care., pp. 175–183 (1979)
16. Shwe, M.A., Middleton, B., Heckerman, D.E., Henrion, M., Horwitz, E.J., Lehnmann, H.P., Cooper, G.F.: Probabilistic diagnosis using a reformulation of the INTERNIST-1/QMP knowledge base. Methods Inf. Med. **30**, 241–255 (1991)
17. Isoc, D.: Faults, diagnosis, and fault detecting structures in complex systems, In: Study and Control of Corrosion in the Perspective of Sustainable Development of Urban Distribution Grids—The 2nd International Conference, Miercurea Ciuc, Romania, June 19–21, pp. 5–12 (2003)

Towards ICT Revolution in Healthcare: Present and Perspectives for Electronic Healthcare Record Systems

Iulian Furdu and Bogdan Pătruţ

Abstract This chapter will describe and discuss the applications and solutions under development or implemented in the e-Health care systems, from the technological, social, organizational dimensions. A survey of the present status in relation with e-Government will cover the leading countries (and not only) in ICT-based developments in these sectors. The major implemented solutions will be outlined regarding their actual implementation and administration. Key aspects will be outlined for Electronic Healthcare Record Systems as core systems in present or future national or regional health programs.

1 Background

Healthcare regardless of it's geographical location and sociopolitical environment, can be viewed as consisting of three stakeholder groups (providers, supporting industry and governance) sharing the common aim of providing the best services to a fourth stakeholder group—the patients, as beneficiaries of these services. Each of these stakeholders will have shared values, expectations, needs, and challenges, which will finally form the growth drivers—or opposers—to the enablement of a common practice in these services. Cost savings, improved patient safety and improved access to care are made possible through ICT investment in areas which have a well-proven business case, such as Electronic Transfer of Prescriptions, Clinical Decision Support, Electronic Health Records, and Chronic Disease Management Systems, all sustained by a modern IT infrastructure. As world's population is ageing healthcare system should support the current and future needs of the population it serves.

I. Furdu (✉) · B. Pătruţ
Vasile Alecsandri University of Bacău, Bacău, Romania
e-mail: ifurdu@ub.ro

B. Pătruţ
e-mail: bogdan@edusoft.ro

B. Iantovics and R. Kountchev (eds.), *Advanced Intelligent Computational*   65
Technologies and Decision Support Systems, Studies in Computational Intelligence 486,
DOI: 10.1007/978-3-319-00467-9_6, © Springer International Publishing Switzerland 2014

The information provided in this chapter covers mainly the last decade focusing on last years. Current status, problems, solutions, advantages or disadvantages, future trends or new advances are described for Electronic Healthcare Record Systems.

2 Electronic Healthcare Record Systems

Electronic medical records (EMR) manage the clinical operations of healthcare providers and lie at the center of any computerized health information system. Without EMR other systems such as decision support systems cannot be effectively integrated into routine clinical workflow. The multi-provider, multi-specialty interoperable, multi-discipline computerized medical record, which has been a goal for healthcare professionals, administrators and many politicians for the past two decades, is about to become reality in many western countries.

Electronic medical records will tie together a patient's health information from numerous systems (like clinics, doctor's offices, hospitals or pharmacies) to provide one coherent record in a structured format. That way, anyone who has authorization to view it, gets a complete and accurate picture of the health status of the patient.

Related terms to EMR that can be used both interchangeably and generically include electronic health record (EHR), electronic patient record (EPR), computer-based patient record (CPR) etc. [1].

An EMR generally contain a whole range of data in comprehensive or summary form, including the patient past medical history, personal stats, medication and allergies, physical assessment, physical examination, daily charting, nursing care plan, referral, symptoms, diagnoses, treatment, laboratory test results, radiology images, procedures, discharge, diaries, immunization status etc.

Worldwide, in North America, Europe, Asia/Pacific, Australia and New Zealand sustainable efforts are made to implement integrated EMR systems each zonal entity usually having its own approach. Important steps to implement national EMR systems are also taken in developing countries such as from Eastern Europe (Poland, Czech Republic, Slovakia, Romania etc.). For example projects over 40 million euro have been recently granted from EU grant for the implementation in Romania of the electronic prescription and electronic patient data sheet with deadlines 2011 and 2012, respectively [2].

3 Problems Identified, Solutions and Current Status

Due to the lack of standardization for EHR systems in US in early 2000s, the amount and the quality of stored data largely depended on software implementation: some of them included virtually all patient data, while others were limited to specific types of data, such as ancillary results and medications. While some EHR systems provided decision support (e.g., practitioner reminders and alerts, alerts concerning possible drug interactions), others did not. Also, most EHR systems were enterprise-specific and few of them provided powerful support for communication and or interconnectivity across the providers within a community [3].

One solution was the adoption of a standard according to US Institute of Medicine report [3] based on eight core capabilities that every EHRs should possess.

Health Level 7 (HL7) version 3 for message transferring and Integrating the Healthcare Enterprise (IHE, www.ihe.net) is a standardized approach to sharing of clinical documents that became the principal messaging standard for clinical data in the U.S., and possibly, in the world.

With the most of the standardization problems resolved, the top remaining problem is the cost of implementation: a full EHR system, including a picture archiving and communication system (PACS) can cost tens of millions, especially if future upgrades are expected. Other technology problems include: the lack of a standard code of generally accepted practices and protocols, poor user interface design, lack of appropriate vocabulary and data transmission standards, difficulty in creating a migration plan from chapter to EHRs. Training costs have to be added: many practitioners are not accustomed to creating and using electronic care records and have to have the literacy in using the system, performance data entry as well as information retrieval. Even implementing an EHR system is a significant undertaking for any healthcare organization, most of them include among their highest priorities the goal of compliant evaluation and management (E/M) coding. [4]: "problems arise from software systems that (a) may have coding engines that fail to account for medical necessity; (b) may have designs that automatically guide physicians to create records with high levels of documented care for every visit; (c) may have shortcut documentation tools that create "automated" documents, identified by Department of Health and Human Services (HHS) as "having the potential for fraud and abuse"; and (d) therefore consistently derive and recommend submission of high-level E/M codes for almost every patient encounter". The analysis of [5] shows that similar critical areas exist in the various countries. Strategic, organizational and human challenges are usually more difficult to master than technical aspects.

The main challenges to implement a national EHR system remain: data transfer is expensive and difficult, lack of common standards for interoperability or data security, and lack of national IT architecture.

The Obama administration has promised to invest $10 billion per year over the next 5 years on healthcare IT. The administration's stimulus package provides incentives for implementing certified EHR systems, while those practices that don't adopt these systems by 2014 will receive reductions in reimbursement.

In Canada, Infoway is an independent, not-for-profit organization that invests in partnership with public sector to implement health information systems. According to its 2006–2007 Annual Report [6] "EHR... at the crossroads of success" the goal for Infoway is that "by 2010 every province and territory and the populations they serve will benefit from new health information systems that will help modernize the healthcare system. Further, 50 % of Canadians will have their electronic health record readily available to the authorized professionals who provide their healthcare services". Canada's successes include the development of common architecture and national standards for interoperability, patient registries and the deployment of digital imaging while permitting local regions to implement their own ICT systems such as SWODIN- the Southwestern Ontario Diagnostic Imaging Network. Criticism to Infoway's large centralized technology systems claims that the focus should have been to start with primary care and add inter-operability at a later stage [7].

World's first digital national EHR belongs to Finland: "Finland is the first country worldwide to offer such an innovative healthcare service to its population." (Anne Kallio, Development director, Ministry of Social Affairs and Health, Finland).

epSOS (www.epsos.eu) is the first pan-European project that provides cross-border interoperability of EHRs between European countries. Guidelines on eHealth interoperability (European Commission's publication in 2007) were a step before. Pilot eHealth infrastructure projects at the national level to implement interopera-bility standards and architecture that are compatible with HL7 are developed in England, Wales and Denmark (which has its own standard for messaging named MedCon), Sweden by Carelink: national organisation that co-ordinates the devel-opment and use of IT by medical professionals. Carelink also manages Sjunet, the Swedish national IT infrastructure for healthcare. In France the system SESAM-Vitale electronic health insurance (actually at its second generation called Vitale 2) uses two cards to electronically sign claim forms—the patient data card and a health professional card in the same reader—and the forms are sent directly to the patient's health insurance provider. According to Gemalto provider (http://www.gemalto. com/public_sector/healthcare/france.html) over 83.63 % of general practitioners are using the system, 99.65 % of pharmacists, 81 % of dentists (July 2010).

Also in Asia/Pacific area, Singapore government has announced that, the state will have a national EHR system by 2010 which is currently in the implementation stage. For the same year in Hong Kong was scheduled a pilot program for easier patient registration with the use of Smart ID Cards in hospitals and clinics, linked to clinical records. A major milestone in Taiwan's health industry was the successfully implemented "Smart Card" for healthcare. In New South Wales, Australia, one of the largest clinical information system implementations started in 2007. It implies developing and implementing EMR on a statewide basis, which include the Elec-tronic Discharge Referral System. Also, New Zealand benefits from one of the highest rates of EMR adoption.

4 Benefits

Although its fundament is clinical information, the EHR is used except clinicians by potentially every other health professional who manages healthcare quality, payment, risk, research, education, and operations. Furthermore, EHR provide ubiquitous access, complete and accurate documentation of all clinical details and variances in treatments, interfaces with labs, registries, more reliable prescribing, complete documentation that facilitates accurate coding and billing. Care delivery to remote or rural regions is improved and redundant tests and treatments are eliminated. Doctors and healthcare organizations using fully implemented EHR report lower cost and higher productivity. The study presented in [8] reveals that EHRs help to avert costs and increase revenue leading to significant savings for the healthcare practice in: drug expenditures (33 %), improved utilization of radiology tests (17 %), better capture of charges (15 %), decreased billing errors (15 %). Patients, too, benefit from increasingly gaining access to their health information and making important contributions to their personal health. Electronic referrals allow for easier access to follow-up care with specialists, patients have portals for online interaction with various providers; the need to fill out repeatedly same forms is reduced. EHRs serve as the foundation for population health and, ultimately, the potential for a national health information infrastructure.

The most widely EHR systems implemented are England, Denmark, Netherlands, and certain regions of Spain which are close to 100 % regarding to the use of ambulatory EHR. Also, Sweden, Norway are at 80 % and behind Germany/France are at 50 %. US is somewhere less 20 %, depending on EHR classification [9]. From the perspective of EHR using for inpatient, coverage is high in England, Sweden, Norway, Denmark, and Finland, followed by Germany and Spain (in mid-low tier hospitals). In the US, Computerized physician order entry CPOE adoption nationally is less than 25 % [9]. If interoperability is to consider Denmark has the most signification implementation of production EHR with over 90 % of encounters shared electronically same as some regions in Spain, English, and Sweden.

5 Conclusions

The entire world is facing a healthcare revolution: the need for better solutions according to ever growing needs for high quality services has driven health care institutions, government policies, software or hardware providers to embrace or to adapt the most valuable solution: e-Health. New national e-Health systems will be functional next years and the implemented ones will be in a constant development. The corresponding market for such services most probably will register a solid growth next years.

Acknowledgments This work was supported by the Bilateral Cooperation Research Project between Bulgaria and Romania (2010–2012) entitled *"Electronic Health Records* for the Next Generation Medical Decision Support in Romanian and Bulgarian National Healthcare Systems", NextGenElectroMedSupport.

References

1. Amatayakul, M.K.: Electronic Health Records. A Practical Guide for Professionals and Organizations, 4th edn. American Health Information Management Association (2009)
2. Vasilache, A.: Peste 40 de milioane de euro, fonduri UE nerambursabile, pentru implementarea in Romania a retetei electronice si a fisei electronice a pacientului (in Romanian). Retrieved 15 May 2011, from http://economie.hotnews.ro/stiri-telecom-8314828-peste-40-milioane-euro-fonduri-nerambursabile-pentru-implementarea-romania-retetei-electronice-fisei-electronice-pacientului.htm (2011)
3. Tang, P.: Key Capabilities of an Electronic Health Record System: Letter Report. Committee on Data Standards for Patient Safety. Board on Health Care Services. Institute of Medicine. National Academies Press, Washington DC (2003)
4. Grider, D., Linker, R., Thurston, S., Levinson, S.: The problem with EHRs and coding, Med. Econ. Retrieved 01 May 2011, from http://www.modernmedicine.com/modernmedicine/article/articleDetail.jsp?id=590411 (2009)
5. Deutsch, E., Duftschmida, G., Dorda, W.: Critical areas of national electronic health record programs-Is our focus correct? Int. J. Med. Inf. 79(3), 211–222 (2010)
6. EHR… at the crossroads of success. Annual report. Canada Health Infoway. Retrieved 15 April 2011, from https://www2.infoway-inforoute.ca/Documents/AnnualReport0607-E.pdf (2006–2007)
7. Webster, P., Kondo, W.: Medical data debates: big is better? Small is beautiful? CMAJ 183(5), 539–540 (2011)
8. Wang, S., Middleton, B., Prosser, L., Bardon, C., Spurr, C., Carchidi, P., Kittler, A., Goldszer, R., Fairchild, D., Sussman, A., Kuperman, G., Bates, D.: A cost-benefit analysis of electronic medical records in primary care. AJM 114, 397–403 (2003)
9. Halamka, J.: International EHR Adoption. Healthcare IT News. Retrieved 25 April 2011, from http://www.healthcareitnews.com/blog/international-ehr-adoption (2009)

Compression of CT Images with Branched Inverse Pyramidal Decomposition

Ivo R. Draganov, Roumen K. Kountchev and Veska M. Georgieva

Abstract In this chapter a new approach is suggested for compression of CT images with branched inverse pyramidal decomposition. A packet of CT images is analyzed and the correlation between each couple inside it is found. Then the packet is split into groups of images with almost even correlation, typically into six or more. One is chosen as a referent being mostly correlated with all of the others. From the rest difference images with the referent are found. After the pyramidal decomposition a packet of spectral coefficients is formed and difference levels which are coded by entropy coder. Scalable high compression is achieved at higher image quality in comparison to that of the JPEG2000 coder. The proposed approach is considered perspective also for compression of MRI images.

Keywords CT image · Compression · Branched inverse pyramidal decomposition

1 Introduction

Important stage in Computed Tomography (CT) is archiving the images obtained in an efficient manner concerning the data volume occupied and the image quality. A vast number of medical image compression techniques exist [1] which can be divided into two large groups—lossless [2, 3] and lossy [4] depending on the

I. R. Draganov (✉) · R. K. Kountchev · V. M. Georgieva
Radio Communications and Video Technologies Department, Technical University of Sofia,
8 Kliment Ohridski Blvd, 1000 Sofia, Bulgaria
e-mail: idraganov@tu-sofia.bg

R. K. Kountchev
e-mail: rkountch@tu-sofia.bg

V. M. Georgieva
e-mail: vesg@tu-sofia.bg

B. Iantovics and R. Kountchev (eds.), *Advanced Intelligent Computational*
Technologies and Decision Support Systems, Studies in Computational Intelligence 486,
DOI: 10.1007/978-3-319-00467-9_7, © Springer International Publishing Switzerland 2014

ability to restore the image fully or not. In both groups often a certain type of decomposition of the image is applied—either a linear orthogonal transform or a wavelet one combined with spectral coefficients rearrangement and entropy coding. Some authors propose completely different methods such as the min–max method developed by Karadimitriou and Tyler [5].

Wu [6] propose an approach based on adaptive sampling of DCT coefficients achieving compression in the interval 0.18–0.25 bpp at PSNR between 41 and 43 dB. The quality of the images at this compression is comparable to that of the JPEG2000 as the author shows while the JPEG coder produces images with PSNR between 31 and 40 dB for the same levels. Erickson et al. [7] confirm that wavelet decomposition assures better quality for the images being compressed from 0.1 to 0.4 bpp in comparison to the JPEG coder.

Further more authors undertake the advantages of the wavelet decomposition for medical image compression combining it with other techniques to construct more efficient coders—using joint statistical characterization [8], by linear prediction of the spectral coefficients [9], introducing region of interest (ROI) [10], incorporating planar coding [11], etc. Nevertheless, the higher compression levels achieved some authors point out the significant reduction of the visual quality of these images [4]. While cumulative quality measures such as PSNR stay high the smoothing of vast image areas due to the wavelet coefficients quantization becomes intolerable for compression ratios (CR) smaller than 0.8 bpp in some cases.

In this chapter a new approach for lossy CT image compression is suggested. It is based on linear orthogonal transforms with a new type of spectral coefficients hierarchical grouping provided by the Branched Inverse Pyramidal Decomposition (BIDP). Along with entropy coding the approach assures higher image quality than the previously developed methods at the same CR.

The chapter is arranged as follows: in Sect. 2 are given the steps of the proposed algorithm; in Sect. 3 some experimental results are presented, and then a conclusion is made.

2 Compression of CT Images with BIDP

A new opportunity for achieving highly effective compression of CT images is the usage of BIDP with 3 levels based on orthogonal transforms. It represents a generalization of the Inverse Pyramidal Decomposition (IDP) [12] related to group of CT images.

The new approach called BIDP includes the following stages:

1. Selection of a referent image from the group of CT images based on correlation analysis. For the purpose the correlation coefficient ρ_{xy} should be found between the vectors $\vec{X} = [x_1, x_2, \ldots, x_S]^t$ and $\vec{Y} = [y_1, y_2, \ldots, y_S]^t$ describing the intensity of the pixels inside a couple of images from the group:

$$\rho_{x,y} = \sum_{s=1}^{S}(x_s - \bar{x})(y_s - \bar{y}) \bigg/ \left[\sqrt{\sum_{s=1}^{S}(x_s - \bar{x})^2}\right] \times \left[\sqrt{\sum_{s=1}^{S}(y_s - \bar{y})^2}\right]. \quad (1)$$

Here $\bar{x} = \frac{1}{S}\sum_{s=1}^{S} x_s$ and $\bar{y} = \frac{1}{S}\sum_{s=1}^{S} y_s$ are the average values of the elements x_s and y_s of the both vectors and S is the number of pixels in the images. The selection of referent image is done after calculation of all correlation coefficients for all couples possible from the group of CT images. The number of consecutive images N forming a group for compression from all the images in the CT packet is found according the relation $\text{var}(\rho_{xg,yg}) > \rho_{xd,yd}$ where $\rho_{xg,yg}$ is the correlation coefficient between all the couples of images in the group and the $\rho_{xd,yd}$ is the correlation coefficient between the referent image from the group and the most distant one from the CT packet. As shown in Sect. 3 significant variation exist for the correlation coefficient inside the selected group and outside it there is saturation for its value indicating the limits of the group itself. For a group of N images the number L of all couples $l(p, q)$ is:

$$L = \sum_{p=1}^{N-1} \sum_{q=p+1}^{N} 1(p,q). \quad (2)$$

After calculating all L correlation coefficients ρ_{pq} the index p_0 is found for which it is true that:

$$\sum_{q=1}^{N} \rho_{p_0 q} \geq \sum_{q=1}^{N} \rho_{pq} \text{ for } p, q = 1, 2, \ldots, N, \text{ when } p \neq q \text{ and } p \neq p_0. \quad (3)$$

Then the consecutive number of the referent image for the group is p_0, that is $[B_R] = [B_{p_0}]$.

2. The matrix of the referent image R is divided to blocks with dimensions $2^n \times 2^n$ and each of them is presented with Inverse Pyramidal Decomposition (IDP) with 3 levels:

$$[B_R(2^n)] = [\tilde{B}_{0R}(2^n)] + \sum_{p=1}^{2} [\tilde{E}_{p-1,R}(2^n)] + [E_{2,R}(2^n)], \quad (4)$$

where $[E_{2,R}(2^n)]$ is the matrix of the residual from the decomposition. In the last expression each matrix is with dimensions $2^n \times 2^n$. The first component $[\tilde{B}_{0R}(2^n)]$ for the level $p = 0$ is a rough approximation of the block $[B_R(2^n)]$. It is obtained by applying inverse 2D-DCT over the transformed block $[\tilde{S}_{0R}(2^n)]$ in correspondence with the expression:

$$[\tilde{B}_{0R}(2^n)] = [T_0(2^n)]^{-1}[\tilde{S}_{0R}(2^n)][T_0(2^n)]^{-1}, \quad (5)$$

where $[T_0(2^n)]^{-1}$ is a matrix with dimensions $2^n \times 2^n$ for the inverse 2D-DCT.

The matrix $[\tilde{S}_{0R}(2^n)] = [m_0(u,v).s_{0R}(u,v)]$ is the transform block of the cut 2D-DCT over $[B_R(2^n)]$. Here $m_0(u,v)$ are the elements of the binary matrix-mask $[M_0(2^n)]$ with the help of which the preserved coefficients are being determined $[\tilde{S}_{0R}(2^n)]$ in accordance to the equation:

$$m_0(u,v) = \begin{cases} 1, & \text{if } s_{0R}(u,v) \text{ is preserved coefficient,} \\ 0, & \text{otherwise,} \end{cases} \quad \text{for } u,v = 0,1,\ldots,2^n - 1. \tag{6}$$

The values of the elements $m_0(u,v)$ are chosen by the condition the preserved coefficients $\tilde{s}_{0R}(u,v) = m_0(u,v).s_{0R}(u,v)$ to correspond to those with the highest average energy into the transformed blocks $[S_{0R}(2^n)]$ for all the blocks to which the image has been divided. The transformed block $[S_{0R}(2^n)]$ from $[B_R(2^n)]$ is found by the 2D-DCT:

$$[S_{0R}(2^n)] = [T_0(2^n)][B_R(2^n)][T_0(2^n)], \tag{7}$$

where $[T_0(2^n)]$ is a matrix with dimensions $2^n \times 2^n$ for level $p = 0$ which is used for implementing the DCT.

The rest components in decomposition (4) are the approximation matrices $[\tilde{E}_{p-1,R}(2^{n-p})]$ for $p = 1, 2$. Each of them consists of sub-matrices $[\tilde{E}_{p-1,R}^{k_p}(2^{n-p})]$ with dimensions $2^{n-p} \times 2^{n-p}$ for $k_p = 1, 2,\ldots,4^p$ obtained by its quad-tree split. On the other hand each sub-matrix $[\tilde{E}_{p-1,R}^{k_p}(2^{n-p})]$ is calculated by:

$$[\tilde{E}_{p-1,R}^{k_p}(2^{n-p})] = [T_p(2^{n-p})]^{-1}[\tilde{S}_{pR}^{k_p}(2^{n-p})][T_p(2^{n-p})]^{-1} \text{ for } k_p = 1,2,\ldots,4^p, \tag{8}$$

where 4^p is the number of the branches of the quad-tree in level p of the decomposition; $[T_p(2^{n-p})]^{-1}$—matrix for inverse 2D-WHT; $[\tilde{S}_{pR}^{k_p}(2^{n-p})]$—the transformed block of the cut 2D-WHT of the difference matrix $[E_{p-1,R}^{k_p}(2^{n-p})]$. The elements $\tilde{s}_{pR}^{k_p}(u,v) = m_p(u,v).s_{pR}^{k_p}(u,v)$ of the matrix $[\tilde{S}_{pR}^{k_p}(2^{n-p})]$ depend on the elements $m_p(u,v)$ of the binary mask $[M_p(2^{n-p})]$:

$$m_p(u,v) = \begin{cases} 1, & \text{if } s_{pR}^{k_p}(u,v) - \text{preserved coefficient,} \\ 0 & \text{otherwise.} \end{cases} \quad \text{for } u,v = 0,1,\ldots,2^{n-p} - 1. \tag{9}$$

Here $s_{pR}^{k_p}(u,v)$ are elements of the transformed block $[S_{pR}^{k_p}(2^{n-p})]$ which is obtained by the 2D-WHT:

$$[S_{pR}^{k_p}(2^{n-p})] = [T_p(2^{n-p})][E_{p-1,R}^{k_p}(2^{n-p})][T_p(2^{n-p})]. \tag{10}$$

where $[T_p(2^{n-p})]$ is a matrix with dimensions $2^{n-p} \times 2^{n-p}$ for level $p = 0$ by which WHT is applied.

It is possible to represent each group of four neighbouring elements $\tilde{s}_{pR}^{k_p}(u,v)$ for one and the same u and v in the following way:

$$
\begin{bmatrix}
\tilde{d}_{pR}^{k_p}(u,v) \\
\tilde{d}_{pR}^{k_p+1}(u,v) \\
\tilde{d}_{pR}^{k_p+2}(u,v) \\
\tilde{d}_{pR}^{k_p+3}(u,v)
\end{bmatrix}
= \frac{1}{4}
\begin{bmatrix}
1 & 1 & 1 & 1 \\
0 & 4 & 0 & -4 \\
-4 & 0 & 4 & 0 \\
0 & 0 & -4 & 4
\end{bmatrix}
\begin{bmatrix}
\tilde{s}_{pR}^{k_p}(u,v) \\
\tilde{s}_{pR}^{k_p+1}(u,v) \\
\tilde{s}_{pR}^{k_p+2}(u,v) \\
\tilde{s}_{pR}^{k_p+3}(u,v)
\end{bmatrix},
\tag{11}
$$

which allows to gain even higher correlation between the spectral coefficients since the last three ones for positions $(0,1)$, $(1,0)$ and $(1,1)$ form differences two by two and these differences often are zero valued because neighboring blocks contain almost identical content.

The inverse transform which leads to full restoration of $\tilde{s}_{pR}^{k_p}(u,v)$ is given by:

$$
\begin{bmatrix}
\tilde{s}_{pR}^{k_p}(u,v) \\
\tilde{s}_{pR}^{k_p+1}(u,v) \\
\tilde{s}_{pR}^{k_p+2}(u,v) \\
\tilde{s}_{pR}^{k_p+3}(u,v)
\end{bmatrix}
= \frac{1}{4}
\begin{bmatrix}
4 & -1 & -3 & -2 \\
4 & 3 & 1 & 2 \\
4 & -1 & 1 & -2 \\
4 & -1 & 1 & 2
\end{bmatrix}
\begin{bmatrix}
\tilde{d}_{pR}^{k_p}(u,v) \\
\tilde{d}_{pR}^{k_p+1}(u,v) \\
\tilde{d}_{pR}^{k_p+2}(u,v) \\
\tilde{d}_{pR}^{k_p+3}(u,v)
\end{bmatrix}.
\tag{12}
$$

The difference matrix $[E_{p-1,R}(2^{n-p})]$ for level p containing the sub-matrices $[E_{p-1,R}^{k_p}(2^{n-p})]$ is determined by the following equation:

$$
[E_{p-1,R}(2^{n-p})] = \begin{cases}
[B_R(2^n)] - [\tilde{B}_{0R}(2^n)] & \text{for } p = 1; \\
[E_{p-2,R}(2^{n-p})] - [\tilde{E}_{p-2,R}(2^{n-p})] & \text{for } p = 2.
\end{cases}
\tag{13}
$$

3. Branch is taken only for level $p = 0$ of the pyramid (4) of the referent image R with dimensions $H \times V$. The preserved coefficients $\tilde{s}_{0R}(u,v)$ with all the same spatial frequencies (u,v) from all blocks $[\tilde{B}_{0R}(2^n)]$ for $p = 0$ are united into two-dimensional arrays $[\tilde{S}_{0R}(u,v)]$ with dimensions $(H/2^n) \times (V/2^n)$. The number of these matrices is equal of the number of the preserved coefficients $\tilde{s}_{0R}(u,v)$ in each block from the referent image. In resemblance to Eq. (4) every found matrix $[\tilde{S}_{0R}(u,v)] = [B_{uv}]$ is represented by IPD with 2 levels:

$$
[B_{uv}] = [\tilde{B}_{uv}] + [\tilde{E}_{0,uv}] + [E_{1,uv}],
\tag{14}
$$

where $[E_{1,uv}]$ is the residual from the decomposition. Its components are matrices with dimensions $(H/2^n) \times (V/2^n)$ and they are found in a similar fashion as it was done in (4). The first component for level $p = 0$ is found by:

$$
[\tilde{B}_{uv}] = [T_0]^{-1}[\tilde{S}_0][T_0]^{-1},
\tag{15}
$$

where $[T_0]^{-1}$ is a matrix with dimensions $(H/2^n) \times (V/2^n)$ used for the inverse 2D-WHT. The matrix $[\tilde{S}_0]$ is the transformed block of $[\tilde{B}_{uv}]$ obtained by the cut 2D-WHT:

$$[\tilde{S}_0] = [T_0][\tilde{B}_{uv}][T_0]. \tag{16}$$

The preserved coefficients of the transformed block $[\tilde{S}_0]$ are calculated according to Eq. (6). The next component for $p = 1$ of decomposition (14) is estimated based on the difference:

$$[E_0] = [B_{uv}] - [\tilde{B}_{uv}]. \tag{17}$$

The approximation of this difference is given by:

$$[\tilde{E}_0^{k_1}] = [T_1]^{-1}[\tilde{S}_1^{k_1}][T_1]^{-1} \text{ for } k_1 = 1, 2, 3, 4, \tag{18}$$

where $[\tilde{S}_1^{k_1}]$ is the transformed block returned by the cut 2D-WHT:

$$[S_1^{k_1}] = [T_1][E_0^{k_1}][T_1]. \tag{19}$$

Here $[T_1]$ is a matrix for WHT with dimensions $(H/2n + 1) \times (V/2n + 1)$.

4. For every block of the ith CT image from the group which is not referent a difference is found:

$$[E_{0i}(2^n)] = [B_i(2^n)] - [\tilde{B}_{0R}(2^n)] \text{ for } i = 0, 1, 2, \ldots, N - 1, \tag{20}$$

where N is the number of the CT images in the group.

The difference matrices $[E_{p-1,i}(2^n)]$ for the next levels $p = 1, 2$ are divided to 4^p sub-matrices with dimensions $2^{n-p} \times 2^{n-p}$ and over each one of them is applied the cut 2D-WHT. Further the processing of the obtained matrices is done in a similar way as the processing of the components of the referent image R за $p = 1, 2$.

It should be noticed that when the number of the preserved coefficients in a certain block is 4 using the 2D-WHT it is possible to reduce this number for levels $p = 1, 2$. As shown in [4] for each of these levels it is not necessary to calculate coefficients $s_p^{k_p}(0, 0)$ as they are always zero. Thus, the number of the coefficients necessary for lossless reconstruction of the image becomes smaller with a factor of 1.33.

From the output of the coder the following arrays containing spectral coefficients are passed:

1. From level $p = 0$ of the referent image represented with a branch in the form of pyramid with levels $p = 0, 1$ and residual 3 arrays are formed of coefficients with frequencies (u, v). Then the total amount of arrays is $3 \times$ (number of preserved coefficients) and the length of each array is $(H/2^n) \times (V/2^n)$;
2. From levels $p = 1, 2$ for each of the N-th images in the group are formed arrays of preserved coefficients with frequencies (u, v). The number of the arrays is equal to that of the preserved coefficients and their lengths are equal to $4^p(H/2^{n+p}) \times (V/2^{n+p}) = (H/2^n) \times (V/2^n)$.

Over the coefficients from the output of the coder for all the levels of the branched pyramid for every CT image in the group lossless entropy coding (EC) is applied which includes run-length coding (RLC), Huffman coding (HC) and arithmetic coding (AC).

At the stage of decoding the compressed data for the group of CT images all the operations are carried out in reverse order: lossless decoding, branch matrix restoration based on Eq. (14), referent image decoding according to (4) and the rest images in correspondence to (20). As a result all CT images from the group are restored.

3 Experimental Results

The CT test images are 576 greyscale slices in DICOM format. The size of all images is H = 512 × V = 512 pixels with intensity depth of 16 bpp.

In Fig. 1a the correlation coefficient is presented between each two images from the packet and in Fig. 1b—the same coefficient only between the first image and all the others. As suggested in Sect. 2 a strong variation of the correlation exists inside a candidate group around a proper referent and outside it asymptotically goes to a constant.

The first test group for which experimental results are presented in Table 1 consists from 9 images shown in Fig. 2—the first one appears to be the referent. The size of the initial block is 16 × 16 ($n = 4$) and the number of the preserved coefficients is 7—all of them low-frequent. In the zero and first level of the branch the preserved coefficients are 4—again low-frequent. For the main branch of the inverse pyramid in the first and second level 4 low-frequent coefficients are preserved.

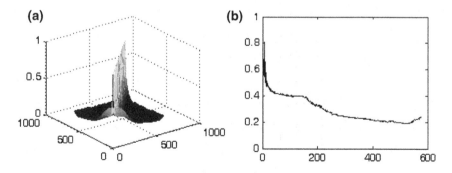

Fig. 1 The correlation coefficient for **a** all couples of images, and **b** the first one and all the others

Table 1 Compression and quality for the group images at three levels of the main pyramid

Image	Level 0			Level 1			Level 2		
	CR, (bpp)	PSNR, (dB)	SSIM	CR, (bpp)	PSNR, (dB)	SSIM	CR, (bpp)	PSNR, (dB)	SSIM
1 (Base)	0.12	38.37	0.9862	0.35	39.84	0.9935	1.33	44.16	0.9955
2 (Diff. 1)	0.10	37.86	0.9867	0.33	39.80	0.9941	1.32	43.38	0.9961
3 (Diff. 2)	0.10	35.53	0.9862	0.33	37.36	0.9937	1.33	43.12	0.9960
4 (Diff. 3)	0.11	35.16	0.9811	0.37	37.11	0.9875	1.44	42.55	0.9871
5 (Diff. 4)	0.11	35.75	0.9706	0.35	36.94	0.9953	1.39	41.98	0.9970
6 (Diff. 5)	0.10	34.12	0.9515	0.34	36.54	0.9650	1.38	41.35	0.9667
7 (Diff. 6)	0.12	33.13	0.9501	0.44	35.27	0.9612	1.74	40.83	0.9669
8 (Diff. 7)	0.12	33.35	0.9487	0.44	35.12	0.9523	1.75	40.56	0.9675
9 (Diff. 8)	0.12	32.01	0.9401	0.43	34.86	0.9594	1.73	40.10	0.9695

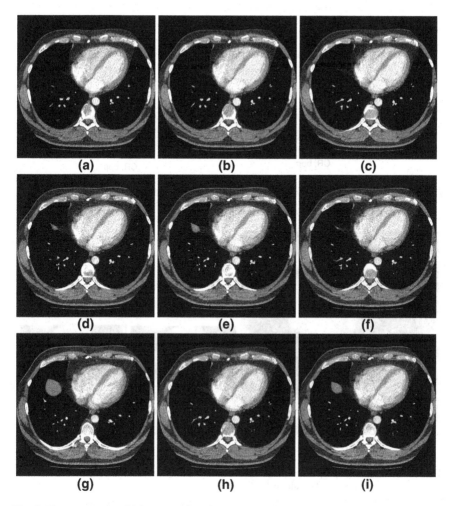

Fig. 2 First test group of 9 images: **a** base image, and **b–i** 8 side images

In Fig. 3 the changes of the average PSNR and the average SSIM from the average CR are given for the group compared to those obtained when JPEG2000 coder is applied over each image separately.

With the exception of the range of low compression (under 1.5 bpp) the suggested approach produces higher value for the average PSNR than JPEG2000. Especially higher is the difference for the big CR values—with 6 dB difference on average. In relation to preserving the structural similarity of the compressed images it is also visible that the proposed approach is dominating JPEG2000—at 0.1 bpp with more than 0.05 for the SSIM.

In Fig. 4a is shown the referent image with an isolated area magnified after compressing with BIDP in Fig. 4b and with JPEG2000 in Fig. 4c at CR = 0.11 bpp.

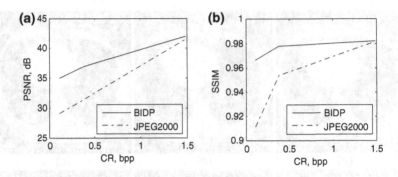

Fig. 3 Quality comparison between BIDP and JPEG2000 based on **a** PSNR, and **b** SSIM

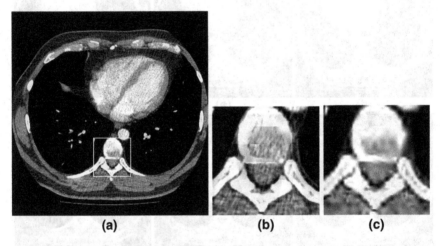

Fig. 4 Visual quality comparison for a segment of the **a** original referent image compressed at CR = 0.11 bpp using, **b** BIDP, and **c** JPEG2000

Worsening the quality for JPEG2000 is obvious in comparison to the BIDP coder—even vast homogenous areas are highly blurred and no details are visible in practice. Only slight block effect is present when applying BIDP at such high CR changing the smaller details insignificantly.

4 Conclusion

From the presented experimental results the advantages of the proposed approach using BIDP become evident when compressing CT images. The coder presented proves to be more efficient than the widely used in practice JPEG2000 coder. Considerably high values for the compression ratio are achieved while preserving high quality of the images—around 39 dB on average and in some cases—over 44 dB. The structural similarity index is close to 1. With the introduction of quantization tables for the spectral coefficients being transmitted it is possible to achieve smooth change for the compression ratio. With the increase of the size of the images it is suitable to increase the size of the initial block (working window, $2^n \times 2^n$) and the number of the levels of the pyramid. Possibility for further development of the proposed approach is applying it over MRI images.

Acknowledgments This chapter was supported by the Joint Research Project Bulgaria-Romania (2010–2012): "Electronic Health Records for the Next Generation Medical Decision Support in Romanian and Bulgarian National Healthcare Systems", DNTS 02/19.

References

1. Graham, R.N.J., Perriss, R.W., Scarsbrook, A.F.: DICOM demystified: a review of digital file formats and their use in radiological practice. Clin. Radiol. **60**, 1133–1140 (2005)
2. Clunie, D.A.: Lossless compression of grayscale medical images: effectiveness of traditional and state of the art approaches. In: Proceedings of SPIE, vol. 3980, pp. 74–84 (2000)
3. Kivijarvi, J., Ojala, T., Kaukoranta, T., Kuba, A., Nyu'l, L., Nevalainen, O.: A comparison of lossless compression methods for medical images. Comput. Med. Imaging Graph. **22**, 323–339 (1998)
4. Ko, J.P., Chang, J., Bomsztyk, E., Babb, J.S., Naidich, D.P., Rusinek, H.: Effect of CT image compression on computer-assisted lung nodule volume measurement. Radiology **237**, 83–88 (2005)
5. Karadimitriou, K., Tyler, J.M.: Min-max compression methods for medical image databases. ACM SIGMOD Rec. **26**, 47–52 (1997)
6. Wu, Y.G.: Medical image compression by sampling DCT coefficients. IEEE Trans. Inf. Technol. Biomed. **6**(1), 86–94 (2002)
7. Erickson, B.J., Manduca, A., Palisson, P., Persons, K.R., Earnest, F., Savcenko, V., Hangiandreou, N.J.: Wavelet compression of medical images. Radiology **206**, 599–607 (1998)
8. Buccigrossi, R.W., Simoncelli, E.P.: Image compression via joint statistical characterization in the wavelet domain. IEEE Trans Image Process **8**(12), 1688–1701 (1999)
9. Ramesh, S.M., Shanmugam, D.A.: Medical image compression using wavelet decomposition for prediction method. Int. J. Comput. Sci. Inf. Secur. (IJCSIS) **7**(1), 262–265 (2010)
10. Gokturk, S.B., Tomasi, C., Girod, B., Beaulieu, C.: Medical image compression based on region of interest, with application to colon CT images. In: Proceedings of the 23rd Annual International Conference of the IEEE Engineering in Medicine and Biology Society, vol. 3, pp. 2453–2456 (2001)
11. Lalitha, Y.S., Latte, M.V.: Image compression of MRI image using planar coding. Int. J. Adv. Comput. Sci. Appl. (IJACSA) **2**(7), 23–33 (2011)
12. Kountchev, R.K., Kountcheva, R.A.: Image representation with reduced spectrum pyramid. In: Tsihrintzis, G., Virvou, M., Howlett, R., Jain, L. (eds.) New Directions in Intelligent Interactive Multimedia. Springer, Berlin (2008)

Adaptive Interpolation and Halftoning for Medical Images

Rumen Mironov

Abstract Two methods for local adaptive two-dimensional processing of medical images are developed. In the first one the adaptation is based on the local information from the four neighborhood pixels of the processed image and the interpolation type is changed to zero or bilinear. In the second one the adaptive image halftoning is based on the generalized 2D LMS error-diffusion filter. An analysis of the quality of the processed images is made on the basis of the calculated PSNR, SNR, MSE and the subjective observation. The given experimental results from the simulation in MATLAB 6.5 environment of the developed algorithms, suggest that the effective use of local information contributes to minimize the processing error. The methods are extremely suitable for different types of images (for example: fingerprints, contour images, cartoons, medical signals, etc.). The methods have low computational complexity and are suitable for real-time applications.

Keywords Image interpolation · Local adaptation · Image processing · Image quantization · Error diffusion · Adaptive filtration · LMS adaptation

1 Introduction

The large variety of visualization devices, used for the transmission, processing and saving of video information (monitors, printers, disks and etc. with different resolution and capacity) leads to the necessity of compact and fast algorithms for image scaling and image halftoning.

R. Mironov (✉)
Department of Radio Communications and Video Technologies, Technical University of
Sofia, Boul. Kl. Ohridsky 8, 1000 Sofia, Bulgaria
e-mail: rmironov@tu-sofia.bg

B. Iantovics and R. Kountchev (eds.), *Advanced Intelligent Computational*   83
Technologies and Decision Support Systems, Studies in Computational Intelligence 486,
DOI: 10.1007/978-3-319-00467-9_8, © Springer International Publishing Switzerland 2014

The basic methods for 2D image interpolation are separated in two groups: non-adaptive (zero, bilinear or cubic interpolation) [1–5] and adaptive interpolation [6–17]. A specific characteristic for the non-adaptive methods is that when the interpolation order increases, the brightness transitions sharpness decreases. On the other side, in result of the interpolation order decreasing, artefacts ("false" contours) in the homogeneous areas are depicted. To reduce them more sophisticated adaptive image interpolation methods were proposed in the recent years [13–17], etc. These methods are based on edge patterns prediction in the local area (minimum 4 × 4) and on adaptive high-order (bicubic or spline) interpolation with contour filtration. The main insufficiency of these methods is that the analysis is very complicated and the image processing requires too much time.

The linear filtration is related to the common methods for image processing and is separated into the two basic types—non-adaptive and adaptive [18, 19]. In the second group the filter parameters are obtained by the principles of the optimal (Winner) filtration, which minimizes the mean square error of signal transform and assumes the presence of the a priory information for image statistical model. The model inaccuracy and the calculation complexity required for their description might be avoided by adaptive estimation of image parameters and by iteration minimization of mean-square error of the transform.

Depending on the processing method, the adaptation is divided into global and local. The global adaptation algorithms refer mainly to the basic characteristics of the images, while the local ones are connected to adaptation in each pixel of the processed image based on the selected pixel neighborhood.

In the present chapter two local adaptive image processing algorithms for image halftoning and linear prediction are developed. The coefficients of the filters are adapted with the help of generalized two-dimensional LMS algorithm [20–23].

This work is arranged as follows: Sect. 2 introduces the mathematical description of the new adaptive 2D interpolation; Sect. 3 introduces the mathematical description of the new adaptive 2D error-diffusion filter; Sect. 4 gives some experimental results and in Sect. 5 concludes this chapter.

2 Mathematical Description of Adaptive 2D Interpolation

The input halftone image of size M × N with m-brightness levels and the interpolated output image of size pM × qN can be presented as follows:

$$A_{M \times N} = \{a(i,j)/i = \overline{0, M-1}; j = \overline{0, N-1}\},$$
$$A^*_{pM \times qN} = \{a^*(k,l)/k = \overline{0, pM-1}; l = \overline{0, qN-1}\}, \tag{1}$$

where q and p are the interpolation's coefficients in horizontal and vertical direction [5, 6].

Fig. 1 Generalized block
diagram of the 2D
interpolator

a(i,j)	a(i,j+1)
a(i+1,j)	a(i+1,j+1)

The differences between any two adjacent elements of the image in a local neighborhood of size 2×2, as shown on Fig. 1, can be described by the expressions:

$$\Delta_{2m+1} = |a(i+m,j) - a(i+m,j+1)|, \quad \text{for} \quad m = 0,1\,;$$
$$\Delta_{2n+2} = |a(i,j+n) - a(i+1,j+n)|, \quad \text{for} \quad n = 0,1\,. \tag{2}$$

These image elements are used as supporting statements in image interpolation.

Here are introduced four logic variables f_1, f_2, f_3 and f_4, which depend on the values of the differences of the thresholds for horizontal θ_m and vertical θ_n direction in accordance with the expressions:

$$f_{2m+1} = \begin{cases} 1, \text{if} : \Delta_{2m+1} \geq \theta_m \\ 0, \text{if} : \Delta_{2m+1} < \theta_m \end{cases}; \quad f_{2n+2} = \begin{cases} 1, \text{if} : \Delta_{2n+2} \geq \theta_n \\ 0, \text{if} : \Delta_{2n+2} < \theta_n \end{cases}. \tag{3}$$

Then each element of the interpolated image can be represented as a linear combination of the four supporting elements from the original image:

$$a^*(k,l) = \sum_{m=0}^{1} \sum_{n=0}^{1} w_{m,n}(r,t) a(i+m,j+n) \tag{4}$$

for $r = \overline{0,p}$; $t = \overline{0,q}$. The interpolation coefficients:

$$w_{m,n}(r,t) = F \cdot ZR_{m,n}(r,t) + \overline{F} \cdot BL_{m,n}(r,t) \tag{5}$$

depend on the difference of the logical function F, which specifies the type of interpolation (zero or bilinear): $F = f_1 f_3 \cup f_2 f_4$. The coefficients of the zero (ZR) and the bilinear (BL) interpolation are determined by the following relations:

$$ZR_{m,n}(r,t) = \frac{1}{4}[1 - (-1)^m \text{sign}(2r - p)][1 - (-1)^n \text{sign}(2t - q)]$$
$$BL_{m,n}(r,t) = (-1)^{m+n}\left[1 - m - \frac{r}{p}\right]\left[1 - n - \frac{t}{q}\right] \tag{6}$$

The dependence of function F of the variables f_1, f_2, f_3 and f_4, defining the type of luminance transition in a local window with size 2×2, is shown in Table 1. In the image for homogeneous areas ($F = 0$) the bilinear interpolation is used, and in the non homogeneous areas ($F = 1$)—the zero interpolation is used.

The two-dimensional interpolation process can be characterized by the following generalized block diagram shown in Fig. 2.

Table 1 The dependence of function F of the variables f_1, f_2, f_3 and f_4

	f_1	f_2	f_3	f_4	F	Transitions		f_1	f_2	f_3	f_4	F	Transitions
0	0	0	0	0	0		8	1	0	0	0	0	
1	0	0	0	1	0		9	1	0	0	1	0	
2	0	0	1	0	0		A	1	0	1	0	1	
3	0	0	1	1	0		B	1	0	1	1	1	
4	0	1	0	0	0		C	1	1	0	0	0	
5	0	1	0	1	1		D	1	1	0	1	1	
6	0	1	1	0	0		E	1	1	1	0	1	
7	0	1	1	1	1		F	1	1	1	1	1	

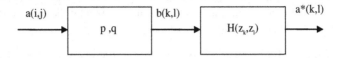

Fig. 2 Structure of the supporting image elements

In the unit for the secondary sampling, the frequencies f_{sr} and f_{sr} were increased p and q times in vertical and horizontal direction. Accordingly, the elements $a(i,j)$ of the input image are complemented with zeros to obtain the elements $b(k,l)$ by the following expression:

$$b(k, l) = \begin{cases} a(k/p, l/q), & \text{for } k = \overline{0, \pm(M - 1)p}, \ l = \overline{0, \pm(N - 1)q}, \\ 0, & \text{otherwise.} \end{cases} \quad (7)$$

The resulting image is processed by a two-dimensional digital filter with transfer function $H(z_k, z_l)$ and the resulting output are the elements $a^*(k,l)$ of the interpolated image. In this case the expression (4) can be presented as follows:

$$a^* (k,l) = \sum_{m=0}^{1} \sum_{n=0}^{1} w_{m,n} (r,t).b \left(\left[\frac{k}{p} \right] + pm, \left[\frac{l}{q} \right] + qn \right) \quad (8)$$

where with the operation $[x]$ it's indicated the greatest integer not exceeding x.

Since the interpolation coefficients are repeated periodically, the analysis can be performed on one block of the image, as shown in Fig. 3. With the red line are marked the values for the output image elements by the bilinear interpolation and with the green—the corresponding values at zero interpolation. With black arrows are marked the four supporting image elements in the input image.

Then the relationship between image elements from the input block and output block can be represented as follows:

$$y(r,t) = \sum_{m=0}^{1} \sum_{n=0}^{1} w_{m,n} (r,t).x (pm, qn) \quad (9)$$

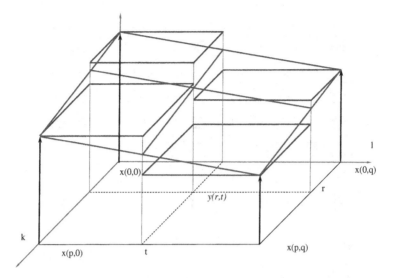

Fig. 3 2D interpolation scheme for one block of the image

where x(pm, qn) are the supporting elements in the current block b(k,l) and y(r,t) are the interpolated elements from the output block a*(k,l).

2.1 Digital Scaling of Halftone Images

The image scaling can be described as a sequence of integer interpolation and decimation. This process can be illustrated with arbitrary fragment from the input image with size 3×3, including 9 image elements—a(i,j), a(i,j+1), a(i,j+2), a(i+1,j), a(i+1,j+1), a(i+1,j+2), a(i+2,j), a(i+2,j+1), a(i+2,j+2). They are used as supporting points in the interpolated image with selected scaling factors—n = 5/2 and m = 3/2. After decimation the output image elements are chooses—b(k,l). The calculation time can be vastly reduced if at the interpolation only output elements are calculated. Every output element b(k,l) is received from four supporting points—a(i,j), a(i,j+1), a(i+1,j), a(i+1,j+1) by the following calculation of corresponding indices:

$$j = \left[\frac{l.px}{px}\right] \text{- for the horizontal direction and } i = \left[\frac{k.qy}{py}\right] \text{- for the vertical direction,}$$

where: m = py/qy, n = px/qx; px, qx, py, qy—are integer; the operation $[X]$ is the integer part of X. In Fig. 4 the distribution of input and output image elements in the scaling image fragment is given. By the scaling, the equations for calculation of output image elements for bilinear and zero interpolation are modified as is shown in the next equations:

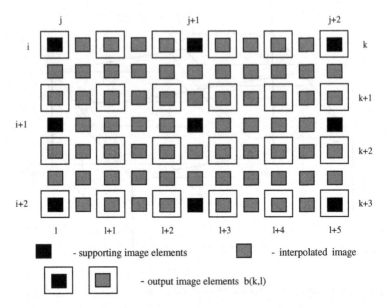

Fig. 4 Distribution of input and output image elements in the scaling fragment

$$b(k,l) = \sum_{m=0}^{1}\sum_{n=0}^{1}(-1)^{m+n}\left[1 - m - \frac{ay}{py}\right]\left[1 - n - \frac{ax}{px}\right].a\,(i + m, j + n) \qquad (10)$$

$$b(k,l) = \sum_{m=0}^{1}\sum_{n=0}^{1}\frac{1}{4}[1 - (-1)^{m}\text{sign}(2ay - py)][1 - (-1)^{n}\text{sign}(2ax - px)].$$
$$a\,(i + m, j + n) \qquad (11)$$

where $ax \equiv (qx.j)\text{mod}(px)$ and $ay \equiv (qy.i)\text{mod}(py)$.

The described scaling algorithm is used for the experiments with the developed adaptive interpolation module.

3 Mathematical Description of Adaptive 2D Error-Diffusion

The input m-level halftone image and the output n-level ($2 \leq n \leq m/2$) image of dimensions $M \times N$ can be represented by the matrices:

$$C = \left\{c(k,l)/k = \overline{0, M-1}; \quad l = \overline{0, N-1}\right\},$$
$$D = \left\{d(k,l)/k = \overline{0, M-1}; \quad l = \overline{0, N-1}\right\}. \qquad (12)$$

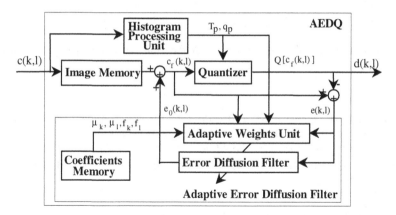

Fig. 5 Adaptive 2D error-diffusion quantizer

Transformation of the image elements $c(k,l)$ into $d(k,l)$ is accomplished by the adaptive error diffusion quantizer (AEDQ) shown on Fig. 5.

The quantizer operation is described by the following equation:

$$d(k,l) = Q[c_f(k,l)] = \begin{cases} q_0, & \text{if } c_f(k,l) < T_0 \\ q_p, & \text{if } T_{p-1} < c_f(k,l) < T_p \quad (p = \overline{1, n-2}). \\ q_{n-1}, & \text{if } c_f(k,l) < T_0 \end{cases} \quad (13)$$

where $q_p \leq q_{p+1} \leq m$ $(p = \overline{0, n-2})$ are the values of the function $Q[.]$.

Thresholds for comparison are calculated by the equation $T_p = (C_p + C_{p+1})/2$, where C_p represents the gray values dividing the normalized histogram of the input halftone image C into n equal parts. The value of the filtered element $c_f(k,l)$ in Eq. (13) is:

$$c_f(k,l) = c(k,l) + e_0(k,l) \quad (14)$$

The summarized error can be expressed as:

$$e_0(k,l) = \sum_{(r,t) \in W} \sum w_{k,l}(r,t)\, e(k-r, l-t) = W_{k,l}^t E_{k,l} \quad (15)$$

where $e(k,l) = c_f(k,l) - d(k,l)$ is the error of the current filtered element when its value is substituted by q_p; $w_{k,l}(r,t)$ are the filter weights defined in the certain causal two-dimensional window W; $W_{k,l}$ and $E_{k,l}$ are the vectors of the weights and their summarized errors, respectively.

According to 2D-LMS algorithm [7] the adaptive error diffusion filter (AEDF) weights can be determined recursively:

$$W_{k,l} = f_k W_{k,l-1} - \mu_k \nabla_{k,l-1} + f_l W_{k-1,l} - \mu_l \nabla_{k-1,l} \quad (16)$$

where: $\nabla_{k,l-1}$ and $\nabla_{k-1,l}$ are the gradients of the squared errors by the quantization in horizontal and vertical directions; f_k, f_l—coefficients, considering the direction of the adaptation, where: $f_k + f_l = 1$; μ_k, μ_l—adaptation steps in the respective direction.

According to [5] the convergence and the stability of the AEDF adaptation process is given by the following condition:

$$|f_k - \mu_k \lambda_i| + |f_l - \mu_l \lambda_i| < 1 \tag{17}$$

where λ_i are the eigen values of the gray-tone image covariance matrix.

Sequence (16) is 2D LMS algorithm of Widrow summary from which the following two particular cases should hold:

First. If $f_k = 1$, $\mu_k = \mu$, $f_l = \mu_l = 0$ then the adaptive calculation of the weights is preceded only in the horizontal direction:

$$\mathbf{W}_{k,l} = \mathbf{W}_{k,l-1} + \mu_k(-\nabla_{k,l-1}) = \mathbf{W}_{k,l-1} - \mu \frac{\partial e^2\,(k,l-1)}{\partial \mathbf{W}_{k,l-1}} \tag{18}$$

Second. If $f_l = 1$, $\mu_l = \mu$, $f_k = \mu_k = 0$ then the adaptive calculation is preceded only in the vertical direction:

$$\mathbf{W}_{k,l} = \mathbf{W}_{k-1,l} + \mu_l(-_{k-1,l}) = \mathbf{W}_{k-1,l} - \mu \frac{\partial e^2\,(k-1,l)}{\partial \mathbf{W}_{k-1,l}} \tag{19}$$

The derivatives by the quantization error in the respective directions are determined by the Eqs. (12–16). For the derivative in horizontal direction is obtained:

$$\frac{\partial e^2(k,l-1)}{\partial \mathbf{W}_{k,l-1}} = 2e(k,l-1)\,\mathbf{E}_{k,l-1}\left[1 - Q'_{c_f}(k,l-1)\right] \tag{20}$$

where:

$$Q'_{c_f}(k,l-1) = \begin{cases} 0, & \text{if}: \quad c_f(k,l-1) \neq T_p \\ q_{p+1} - q_p, & \text{if}: \quad c_f(k,l-1) = T_p \end{cases}.$$

In the same way for the derivative in the vertical direction is obtained:

$$\frac{\partial e^2(k-1,l)}{\partial \mathbf{W}_{k-1,l}} = 2e(k-1,l)\mathbf{E}_{k-1,l}\left[1 - Q'_{c_f}(k-1,l)\right] \tag{21}$$

For the AIHF weights the condition must be hold:

$$\sum_{(r,t)}\sum_{\in W} \mathbf{w}_{k,l}(r,t) = 1 \tag{22}$$

which guarantees that $e(k,l)$ is not increased or decreased by its passing through the error filter.

On the basis of analysis, made in Eqs. (18)–(22) the sequence for the components of $\mathbf{W}_{k,l}$ is:

$$w_{k,l}(r,t) = f_k w_{k,l-1}(r,t)$$
$$- 2\mu_k e(k,l-1)\, e(k-r,l-t-1)\left[1 - Q'_{c_f}(k,l-1)\right]$$
$$+ f_l\, w_{k-1,l}(r,t)$$
$$- 2\mu_l e(k-1,l)\, e(k-r-1,l-t)\left[1 - Q'_{c_f}(k-1,l)\right]. \tag{23}$$

4 Experimental Results

For the analysis of the interpolation distortions the mean-square error (MSE), normalized mean-square error (NMSE in %), signal to noise ratio (SNR in dB) and peak signal to noise ratio (PSNR in dB) can be used as a criterion. The analysis of the interpolated images quality is made by simulation with MATLAB 6.5 mathematical package. The obtained results for one test image (512 × 512, 256 gray levels, scaling factor 30 %) shown in Fig. 6 are summarized in Table 2, including calculated threshold value, number of homogenous and contour blocks, MSE, NMSE, SNR and PSNR for interpolation with coefficient of amplification 3.

In Fig. 7 the results from the different kind of interpolations on the test image with expanding coefficient 3 are shown. The visualization is made with scaling factor 30 %.

Fig. 6 Original test image with size 512 × 512, 256 gray levels

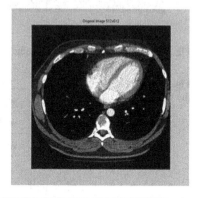

Table 2 Analyse of adaptive interpolation for test image "Spine" with coefficient 3

Image	Spine
$\theta_m = \theta_n$	81
Homogenous blocks	28,361
Contour blocks	200
MSE	47.72
NMSE	6.19×10^{-6}
SNR	52.08
PSNR	31.37

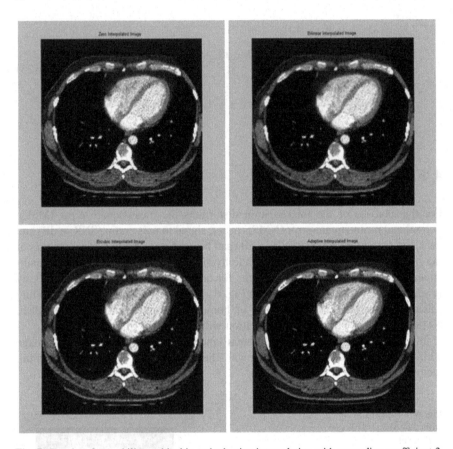

Fig. 7 Results of zero, bilinear, bicubic and adaptive interpolation with expanding coefficient 3

Fig. 8 Spatial disposition of
the weights $w_{k,l}(r,t)$ in W

	l-2	l-1	l	l+1	l+2
k-2	$w^e(.)$	$w^e(.)$	$w^e(.)$	$w^e(.)$	$w^e(.)$
k-1	$w^e(.)$	w(1,1)	w(1,0)	w(1,-1)	
k	$w^e(.)$	w(0,1)	Current		

W^e

W

An error diffusion filter with 4 coefficients has been used for the evaluation of
the efficiency of the developed filter. The spatial disposition, shown on Fig. 8, and
the initial values of weights correspond to these in the Floyd-Steinberg filter [24].

For the calculation of each one of $w_{k,l}(r,t)$ the weights $w^e_{k,l}(.)$ are used from the
extended window W^e.

The analysis of 2D variation of peak signal to noise ratio (PSNR) depending on
parameters f and μ ($f_k = f$, $f_l = 1 - f$, $\mu_k = \mu_l = \mu$.) is made in [22] and [23].

The coefficients f and μ are changed in the following way: $f = 0.0$ to 1.0 with
step 0.1 and $\mu = 1.0 \times 10^{-6}$ to 2.0×10^{-6} with step 1.0×10^{-8}.

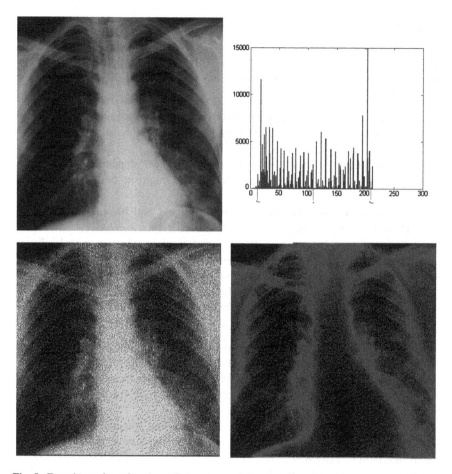

Fig. 9 Experimental results—input X-Ray image "Chest", histogram of image, output bi-level image end error image

The examination of the function PSNR (f, μ) shows that the most proper mean values of f, and μ are f = 0.7, μ = 1.67 × 10⁻⁶. In this case adaptive image halftoning filter leads to the increasing of PSNR with about 0.6 dB in comparison with the 4 coefficient (f = 1, μ = 0) non-adaptive filter of Floyd and Steinberg.

On Fig. 9 the experimental results from the simulation of AEDQ for the test X-Ray image "Chest" with: $M = N = 512$, m = 256 levels, n = 2 bits, $f_k = 0.7$, $f_l = 0.3$, $\mu_k = \mu_l = 1.67 \times 10^{-6}$ are presented.

5 Conclusions

Based on the performed experiments for 2D adaptive interpolation on halftone images, the following conclusions can be made:

- the use of optimal thresholds for selection of homogenous and contour blocks leads to the decreasing of mean-square error, normalized mean-square error and the increasing of signal to noise ratio and peak signal to noise ratio with about 7–10 %;
- by the changing of interpolation type, zero or bilinear, depending on the presence or lack of contours in the area of considerate fragment, better visual quality is achieved;
- the visual quality is better then the zero, bilinear and bicubic interpolation, which is slow for the biggest interpolation coefficients;
- the complexity of the adaptive interpolation is higher than zero and bilinear interpolations but is lower than other high-level interpolations;
- using smaller area for analysis and choice of optimal thresholds for image separation lead to decrease of calculation speed;
- the most effective interpolation for the local characteristic of images can be achieved by using coefficients p, q = 2, 3, 4.

The received results for the quality of interpolated images show that the proposed method for adaptive interpolation can change the high-level interpolations, which are slower in medical systems, using digital image processing and visualization such as digital tomography, archiving of medical information, telemedicine and etc.

The developed generalized adaptive error-diffusion quantizer results in the following particular cases: the wide-spread non-adaptive error diffusion filter of Floyd and Steinberg (for $n = 2$, $f_k = 1$, $\mu_k = \mu_l = f_l = 0$); adaptive error diffusion using the weights only in the horizontal (from the same image row— $f_k = 1$, $f_l = 0$) or only in the vertical direction (from the previous image row— $f_l = 1$, $f_k = 0$). The adaptive filter provides minimum reconstruction error, uniform distribution of the arranged structures in the homogeneous areas and precise reproduction of edges in the output multilevel images. The coefficients f_k, f_l, μ_k, μ_l must be selected on the basis of PSNR analysis and keeping of Eq. (18) as is done in [23]. The developed AEDQ is appropriate for realization on special VLSI circuit to accelerate calculation of image transform and can be used for visualization, printing and transmission of medical images in HD format.

Enhancing the visual quality of the processed medical images through the use of the proposed methods for adaptive image interpolation and halftoning will assist physicians in decision making.

Acknowledgments This work was supported by the Joint Research Project Bulgaria-Romania (2010–2012): *"Electronic Health Records for the Next Generation Medical Decision Support in Romanian and Bulgarian National Healthcare Systems"*, *NextGenElectroMedSupport*.

References

1. Gonzalez, R.C., Woods, R.E.: Digital Image Processing, 3rd edn. Pearson Prentice Hall, New Jersey (2008)
2. Pratt, W.K.: Digital Image Processing. Wiley, New York (2007)
3. Stucki, P.: Advances in Digital Image Processing. Plenum Press, New Jersey (1979)
4. Crochiere, R., Rabiner, L.: Interpolation and decimation of digital signals—a tutorial review. Proc. IEEE **69**(3), 300–331 (1981)
5. Chen, T.C., de Figueiredo, R.P.: Image decimation and interpolation techniques based on frequency domain analysis. IEEE Trans. Commun. **32**(4), 479–484 (1984)
6. Wong, Y., Mitra, S.K.: Edge preserved image zooming, signal processing IV:theories and applications, EURASIP (1988)
7. Kountchev, R., Mironov, R.: Analysis of distortions of adaptive two dimensional interpolation of halftone images, XXV Science Session, "Day of Radio 90", Sofia, Bulgaria, 7–8 May (1990) (in Bulgarian)
8. Kunchev, R., Mironov, R.: Adaptive Interpolation block used in high quality Videowall system. International Conference VIDEOCOMP'90, Varna, Bulgaria, 24–29 Sept (1990)
9. Mironov, R.: Error estimation of adaptive 2D interpolation of images. XXXIX International Scientific Conference on Information, Communication and Energy Systems and Technologies, ICEST'2005, Serbia and Montenegro, Proceedings of Papers vol. 1, pp. 326–329, June 29–July 1 (2005)
10. Mironov, R.: Analysis of quality of adaptive 2D halftone image interpolation. In: Curie, F.J. (ed) National conference with foreign participation, Telecom 2007, International house of scientists "St. Constantine" Resort, Varna, Bulgaria, pp. 117–122, 11–12 Oct (2007)
11. Mironov, R., Kountchev, R.: Optimal thresholds selection for adaptive image interpolation. XLIII International Scientific Conference on Information, Communication and Energy Systems and Technologies, ICEST 2008, Niš, Serbia and Montenegro, pp. 109–112, 25–27 June (2008)
12. Mironov, R., Kountchev, R.: Adaptive contour image interpolation for sign language interpretations. In: The 8th IEEE International Symposium on Signal Processing and Information Technology. ISSPIT 2008, Sarajevo, Bosnia and Herzegovina, pp. 164–169, 16–19 Dec (2008)
13. Carrato, S., Ramponi, G., Marsi, St.: A simple edge-sensitive image interpolation filter. Proc. ICIP'96 (1996)
14. Allebach, J., Wong, P.W.: Edge-directed interpolation. Proc. ICIP'96 (1996)
15. Zhang, X., Wu, X.: Image interpolation by adaptive 2-D autoregressive modeling and soft-decision estimation. IEEE Trans. Image Process **17**(6), 887–896 (2008)
16. Muresan, D.D., Parks, T.W.: Adaptively quadratic (aqua) image interpolation. IEEE Trans. Image Process. **13**(5), 690–698 (2004)
17. Keys, R.G.: Cubic convolution interpolation for digital image processing. IEEE Trans. Acoust. Speech Signal Process. **29**(6), 1153–1160 (1981)
18. Widrow, B., Stearns, S.D.: Adaptive Signal Processing. Prentice-Hall, Inc, New York (1985)
19. Maher, A.: Sid-Ahmed, Image Processing: Theory, Algorithms, and Architectures. McGraw-Hill, Inc, NewYork (1995)
20. Hadhoud, M.M., Thomas, D.W.: The two-dimensional adaptive LMS (TDLMS) algorithm. IEEE Trans. Circuits Syst. **35**(5), 485–494 (1988)
21. Mironov, R., Kunchev, R.: Adaptive error-diffusion method for image quantization, electronics letters, IEE An Intr. Publication the Institution on Electrical Engineers, vol. 29, no. 23, pp. 2021–2023, 11th Nov (1993)
22. Mironov, R.: Algorithms for local adaptive image processing. XXXVII International Scientific Conference on Information, Communication and Energy Systems and Technologies. ICEST 2002, Niš, Yugoslavia, pp. 193–196, 1–4 Oct (2002)

23. Mironov, R.: Analysis of two-dimensional LMS error-diffusion adaptive filter. XXXIX International Scientific Conference on Information, Communication and Energy Systems and Technologies. ICEST 2004, Bitola, Macedonia, pp. 131–134, 16–19 June, (2004)
24. Ulichney, R.A.: Dithering with blue noise. Proc. IEEE **76**(1), 56–79 (1988)
25. Makoto, O., Hashiguchi, S.: Two-dimensional LMS adaptive filters. IEEE Trans. Consum. Electron. **37**(1), 66–73 (1991)

Classification of EEG-Based Brain–Computer Interfaces

Ahmad Taher Azar, Valentina E. Balas and Teodora Olariu

Abstract This chapter demonstrates the development of a brain computer interface (BCI) decision support system for controlling the movement of a wheelchair for neurologically disabled patients using their Electroencephalography (EEG). The subject was able to imagine his/her hand movements during EEG experiment which made EEG oscillations in the signal that could be classified by BCI. The BCI will translate the patient's thoughts into simple wheelchair commands such as "go" and "stop". EEG signals are recorded using 59 scalp electrodes. The acquired signals are artifacts contaminated. These artifacts were removed using blind source separation (BSS) by independent component analysis (ICA) to get artifact-free EEG signal from which certain features are extracted by applying discrete wavelet transformation (DWT). The extracted features were reduced in dimensionality using principal component analysis (PCA). The reduced features were fed to neural networks classifier yielding classification accuracy greater than 95 %.

Keywords Electroencephalography (EEG) · Brain computer interface (BCI) · Decision support system (DSS) · Principal component analysis (PCA) · Independent component analysis (ICA) · Discrete wavelet transformation (DWT) · Artificial neural network (ANN) · Feature extraction · Classification · Computational Intelligence (CI) · Machine learning

A. T. Azar (✉)
Faculty of Engineering, Misr University for Science and Technology (MUST),
6th of October City, Cairo, Egypt
e-mail: Ahmad_t_azar@ieee.org

V. E. Balas
Aurel Vlaicu University of Arad, Arad, Romania
e-mail: balas@drbalas.ro

T. Olariu
Vasile Goldis Western University of Arad, Arad, Romania
e-mail: olariu_teodora@yahoo.com

T. Olariu
Clinical Emergency County Hospital Arad, Arad, Romania

B. Iantovics and R. Kountchev (eds.), *Advanced Intelligent Computational Technologies and Decision Support Systems*, Studies in Computational Intelligence 486, DOI: 10.1007/978-3-319-00467-9_9, © Springer International Publishing Switzerland 2014

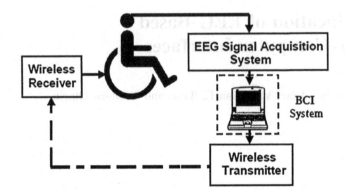

Fig. 1 Block diagram of the system

1 Introduction

Brain computer interface creates a new communication system between the brain and an output device by bypassing conventional motor output pathways of nerves and muscles [1–3]. The BCI operation is based on two adaptive controllers, the user's brain, which produces the activity that encodes the user thoughts, and the system, which decodes this activity into device commands [4, 5]. When advanced Computational Intelligence (CI) and machine learning techniques are used, a Brain computer interface can learn to recognize signals generated by a user after short time of training period [6–14]. The proposed system depends on the EEG activity that derives the user's wishes. Subjects are trained to imagine right and left hand movements during EEG experiment. The imagination of right or left hand movement results in rhythmic oscillations in the EEG signal. The oscillatory activity is comprised of event-related changes in specific frequency bands. This activity can be categorized into event-related desynchronization (ERD), which defines an amplitude (power) decrease of μ rhythm (8–12 Hz) or β rhythm (18–28 Hz), and event-related synchronization (ERS), which characterizes amplitude (power) increase in these EEG rhythms. The system is used to output commands to a remote control to control the movement of a wheelchair via radio frequency (RF) waves (Fig. 1).

2 Methodology

This work is applied on a dataset that uses EEG activity recorded from 59 scalp electrodes according to the international 10/20 system of channel locations. The signals were sampled at 100 Hz. Two different tasks, an imagined right-hand movement and an imagined left-hand movement, are performed in the experiment. The brain signals recorded from the scalp encode information about the user's thoughts. A BCI system has been established to decode this information and translate it into device commands. Figure 2 shows the processing stages of BCI. First the artifacts contaminated in the EEG signal are removed in order to increase the signal to noise ratio (SNR) of the acquired EEG signal. Second, certain features are extracted and translated into device control commands. Each processing phase will be discussed in the following sections.

A. Artifact Removal

Artifacts are non–brain based EEG activity that corrupt and disturb the signal making it unusable and difficult to interpret. To increase the effectiveness of BCI system, it is necessary to find methods for removing the artifacts. The artifact sources can be internal or external. Internal artifacts are those which are generated by the subject itself and uncorrelated to the movement in which we are interested. This type of artifacts includes eye movement, eye blink, heart beat and other muscle activity. On the other hand the external artifacts are coming from the external world such as line noise and electrode displacement. Several approaches for removing these artifacts have been proposed. Early approaches to the task of subtracting artifacts using regression methods were met with limited success [15–17]. Many of the newer approaches involve techniques based on blind source separation. In this chapter, a generally applicable method is applied for removing a wide variety of artifacts based on blind source separation by independent component analysis. The ICA work was performed on Matlab (http://www. mathworks.com) using EEGLAB software toolbox [18]. ICA is a statistical and computational technique that finds a suitable representation of data by finding a suitable transformation. It performs the rotation by minimizing the gaussianity of the data projected on the new axes. By this way it can separate a multivariate signal into additive subcomponents supposing the mutual statistical independence of the non-Gaussian source signals.

Bell and Sejnowski [19] proposed a simple neural network algorithm that blindly separates mixtures, X, of independent sources, S, using information maximization (infomax). They showed that maximizing the joint entropy of the output of a neural processor minimizes the mutual information among the output components. Makeig et al. [16] proposed an approach to the analysis of EEG data based on infomax ICA algorithm. They showed that the ICA can be used to separate the neural activity of muscle and blink artifacts and find the independent

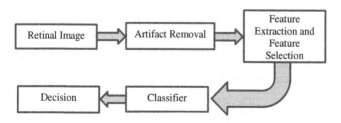

Fig. 2 The processing stages of BCI

components of EEG [20]. Once the independent components are extracted, "corrected EEG" can be derived by identifying the artifactual components and eliminating their contribution to EEG.

ICA methods are based on the assumptions that the signals recorded from the scalp are mixtures of temporally independent cerebral and artifactual sources, that potentials from different parts of the brain, scalp, and the body are summed linearly at the electrodes, and that propagation delays are negligible. ICA will solve the blind source separation problem to recover the independent source signals after they are linearly mixed by an unknown matrix A. Nothing is known about the sources except that there are N recorded mixtures, X. ICA model will be:

$$x = AS \tag{1}$$

The task of ICA is to recover a version U of the original sources, S, by finding a square matrix W, the inverse of matrix A, that invert the mixing process linearly as:

$$U = WX \tag{2}$$

For EEG analysis, the rows of X correspond to the EEG signals recorded at the electrodes, the rows of U correspond to the independent activity of each component (Fig. 3), and the columns of A correspond to the projection strengths of the respective components onto the scalp sensors. The independent sources were visually inspected and artifictual components were rejected to get a "Corrected EEG" matrix, X', by back projection of the the matrix of activation waveforms, U, with artifactual components set to zero, U', as:

$$X' = (W)^{-1} U' \tag{3}$$

Before applying ICA algorithm on the data, it is very useful to do some preprocessing. One popular method is to transform the observed data matrix to obtain a new matrix in which its components are uncorrelated as a condition to be independent. This can be achieved by applying principal component analysis (PCA). PCA finds a transformation for the data to a new orthogonal coordinate

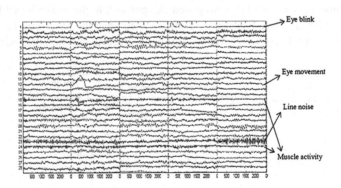

Fig. 3 EEG independent components (ICs)

system with the axes ordered in terms of the amount of variance. At the same time, PCA can be used to reduce the dimension of the data by keeping the principal components that contribute to the most important variance of the data and ignoring the other ones. This often has the effect of reducing the noise and preventing overlearning of ICA.

B. Feature Extraction

To get a reduced and more meaningful representation of the preprocessed signal for further classification, certain features are measured to capture the most important relevant information. The patterns of right and left hand movements are focused in the channels recorded from the sensorimotor area of the brain in the central lobe and some of the other channels might be unusable for discrimination between the two motor tasks. Therefore, a minimum number of EEG channels were selected from the primary sensorimotor cortex area for further processing (Fig. 4). The right cerebral hemisphere of the brain controls the left side of the body and the left hemisphere controls the right side. It was found that left hand movement appears strongly on the C4 channel and right hand movement appears strongly on the C3 channel. The two selected channels were found to be sufficient to ensure a high level of classification as they contain the most relevant information for discrimination [21]. Commonly used techniques for feature extraction such as Fourier analysis have the serious drawback that transitory information is lost in the frequency domain. The investigation of features in the EEG signals requires a detailed time frequency analysis. Wavelet analysis comes into play here since wavelet allows decomposition into frequency components while keeping as much time information as possible [22–26]. Wavelets are able to determine if a

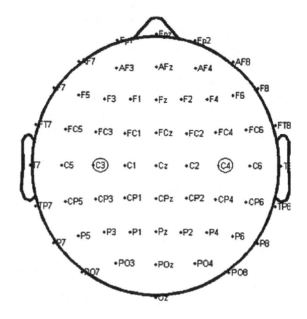

Fig. 4 EEG channel locations showing the selected channels

quick transitory signal exists, and if so, it can localize it. This feature makes wavelets very useful to the study of EEG.

The wavelet transform is achieved by breaking up of a signal into shifted and scaled versions of the original (or mother) wavelet. The mother wavelet ø is scaled by parameter s and translated by τ. WT of a time domain signal x (t) is defined as:

$$w(s, \tau) = \int x(t) \Psi_{s,\tau}^*(t) \, dt \qquad (4)$$

This wavelet transform is called the continuous wavelet transform (CWT). In our case both the input signal and the parameters are discrete so the transform here is the discrete version of wavelet transform. To create the feature vector of each trial, discrete wavelet transform (DWT) was applied. An efficient way to implement DWT is by using digital filter bank using Mallat's algorithm [27].

In this algorithm the original signal passes through two complementary filters, low pass and high pass filters, and wavelet coefficients are quickly produced. This process is iterated to generate at each level of decomposition an approximation cA which is the low frequency component and a detail cD which is the high frequency component (Fig. 5). The ability of the mother wavelet to extract features from the signal is dependent on the appropriate choice of the mother wavelet function. The different orders (wavelets) of the mother wavelet "Coiflet" were tried out to implement the wavelet decomposition (Fig. 6). Each decomposition level corresponds to a breakdown of the main signal to a bandwidth. The low frequency component is the most important part. It carries the information needed about the motor movement found in the ì rhythm (8–12 Hz). Therefore, the coefficients of the second level decomposition cA2 were selected to form the feature vector of each trial. As wavelet coefficients have some redundancy, dimensionality reduction of feature vectors is suggested as a preprocessing step before classification. This could lead to better classification results as it will keep the minimum number of coefficients that are significant and discriminatory. This was achieved by applying PCA by projecting the coefficients onto the first n principal components (PCs), where n is much smaller than the dimensionality of the features. The number of PCs to project the data can be determined by examining the energy of the data. Therefore, PCA was formed by projecting 150 dimensional patterns onto the first 44 PCs which accounted for 99.98 % of the variability of the data.

Fig. 5 Multi-level wavelet decomposition of EEG signal

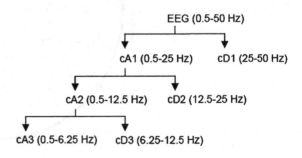

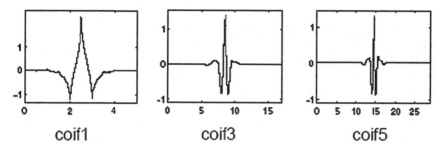

Fig. 6 The mother wavelet (Coiflet) with the different orders used in the decomposition

C. Classification

Classification was performed by training nonlinear feedforward neural networks using the standard backpropagation algorithm to minimize the squared error between the actual and the desired output of the network [28]. The use of nonlinear methods is useful when the data is not linearly separable. The network is used to develop a nonlinear classification boundary between the two classes in feature space in which each decision region corresponds to a specific class.

The network was implemented with 3 hidden neurons in the hidden layer and a single neuron in the output layer that will result in a single value 0 or 1 (Fig. 7). The target output during the training was set to 0 and +1 to represent the different classes. When simulating new input data, an output value greater than or equal to 0.5 represents the first class and a value less than 0.5 represents the other one. The data from one subject was divided into training set and test set using "leave-k-out" cross validation method. By this way the data was divided into k subsets of equal size. The network was trained k times. Each time leaving out one of the subsets from training and using only the remaining subset for validation. To get the true classification rate, the accuracy was averaged over all subsets.

Fig. 7 The architecture of the feed-forward neural network used for classification

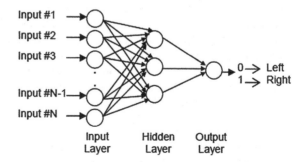

Table 1 Classification accuracy of imagined right and left hand movements for 6 subjects obtained using the different orders of the mother wavelet "Coiflet"

Subject	Coif1 (%)	Coif3 (%)	Coif 5 (%)
Subject 1	94.45	95.56	96.67
Subject 2	92.23	93.34	97.78
Subject 3	91.12	94.45	96.67
Subject 4	90	92.23	97.78
Subject 5	86.67	91.12	100
Subject 6	91.12	94.45	95.56
Average	87.93	93.52	97.41

3 Results and Discussion

To evaluate the performance of our system, EEG data acquired from a motor experiment is processed. The subject made an imagined left or right hand keyboard pressing synchronized with command received. A total of 90 trials were carried out in the experiment for each subject. Table 1 shows the results of all classification experiments as the average percent of test patterns classified correctly using the "leave-k-out" cross validation method.

The results presented above demonstrate that the most effective results were found by applying the mother wavelet "Coiflet" order 5. As there is no well-defined rule for selecting a wavelet basis function in a particular application or analysis, different wavelets were tried out. However, for a more precise choice of a wavelet function, the properties of the wavelet function and the characteristics of the signal to be analyzed should be matched which was the case in the wavelet "Coiflet" of order 5. Also there are other wavelet families can be applied like Harr, Daubechies, and Symmlet. Since the classification accuracy is sensitive to the contaminated EEG artifacts, our processing was performed on artifact-free EEG signal. In order to improve the performance of the system in real time applications, it is ideal if the removal of the artifacts is done using automatic methods [29]. The results also indicated that the two channels C3 and C4 of the sensorimotor cortex area are sufficient to ensure high classification rates. Müller-Gerking et al. [30] and Ramoser et al. [31] studies give a strong evidence that further increase in the number of used channels can increase the classification accuracy. Other studies were performed on the same dataset using a hierarchical multi-method approach based on spatio-temporal pattern analysis achieved classification accuracy up to 81.48 % [32].

4 Conclusions

With the advances in DSS, expert systems (ES) and machine learning, the effects of these tools are used in many application domains and medical field is one of them. Classification systems that are used in medical decision making provide medical data to be examined in shorter time and more detailed. Recently, there has been a great progress in the development of novel computational intelligence techniques for recording and monitoring EEG signal. Brain Computer Interface technology involves monitoring of brain electrical activity using electroencephalogram (EEG) signals and detecting characteristics of EEG patterns by using digital signal processing (DSP) techniques that the user applies to communicate. The proposed wavelet-based processing technique leads to satisfactory classification rates that improve the task of classifying imagined hand movements. The results are promising and show the suitability of the technique used for this application. It is hoped to realize the system in real world environment and to overcome the BCI challenges of accuracy and speed.

Acknowledgments The research of Valentina Emilia Balas was supported by the Bilateral Cooperation Research Project between Bulgaria-Romania (2010-2012) entitled "Electronic Health Records for the Next Generation Medical Decision Support in Romanian and Bulgarian National Healthcare Systems", NextGenElectroMedSupport.

References

1. Wolpaw, J.R.: Brain-computer interfaces as new brain output pathways. J Physiol. **579**(Pt 3), 613–619 (2007)
2. Wolpaw, J.R., Birbaumer, N., McFarland, D.J., et al.: Brain-computer interfaces for communication and control. Clin Neurophysiol. **113**(6), 767–791 (2002)
3. Kübler, A., Kotchoubey, B., Kaiser, J., et al.: Brain-computer communication: unlocking the locked in. Psychol. Bull. **127**(3), 358–375 (2001)
4. Kauhanen, L., Jylanki, P., Janne, L.J. et al.: EEG-based brain-computer interface for tetraplegics. Comput. Intell. Neurosc. l, 11, Article ID 23864, (2007). doi:10.1155/2007/23864
5. Wolpaw, J.R., Birbaumer, N., Heetderks, W.J., et al.: Brain-computer interface technology: A review of the first international meeting. IEEE Trans. Rehab. Eng. **8**(2), 164–173 (2000)
6. Gianfelici, F., Farina, D.: An effective classification framework for brain-computer interfacing based on a combinatoric setting. IEEE Trans. Signal Process. **60**(3), 1446–1459 (2012)
7. Lotte, F., Congedo, M., Lécuyer, A., et al.: Review of classification algorithms for EEG-based brain-computer interfaces. J Neural Eng. **4**(2), R1–R13 (2007)
8. Bashashati, A., Fatourechi, M., Ward, R.K., Birch, G.E.: A survey of signal processing algorithms in brain-computer interfaces based on electrical brain signals. J Neural Eng. **4**(2), R32–57 (2007)
9. Rezaei, S., Tavakolian, K., Nasrabadi, A.M., Setarehdan, S.K.: Different classification techniques considering brain computer interface applications. J Neural Eng. **3**(2), 139–144 (2006)
10. Lemm, S., Blankertz, B., Curio, G., Müller, K.R.: Spatio-spectral filters for improving the classification of single trial EEG. IEEE Trans. Biomed. Eng. **52**(9), 1541–1548 (2005)
11. Dornhege, G., Blankertz, B., Curio, G., Müller, K.R.: Boosting bit rates in noninvasive EEG single-trial classifications by feature combination and multiclass paradigms. IEEE Trans. Biomed. Eng. **51**(6), 993–1002 (2004)
12. Li, Y., Gao, X., Liu, H., Gao, S.: Classification of single-trial electroencephalogram during finger movement. IEEE Trans. Biomed. Eng. **51**(6), 1019–1025 (2004)

13. Sykacek, P., Roberts, S.J., Stokes, M.: Adaptive BCI based on variational Bayesian Kalman filtering: an empirical evaluation. IEEE Trans. Biomed. Eng. **51**(5), 719–727 (2004)
14. del Millán, J., Mouriño, J.R: Asynchronous BCI and local neural classifiers: An overview of the adaptive brain interface project. IEEE Trans. Neural Syst. Rehabil. Eng. **11**(2), 159–61 (2003)
15. Verleger, R., Gasser, T., Möcks, J.: Correction of EOG artifacts in event related potentials of EEG: Aspects of reliability and validity. J. Psychophysiology **19**(4), 472–480 (1982)
16. Makeig, S., Bell, A., Jung, T.P., Sejnowski, T.J.: Independent component analysis of electroencephalographic data. In: Touretzky, D., Mozer, M., Hasselmo, M. (eds.) Advances in neural information processing systems, pp. 145–151 (1996)
17. Jung, T.P., Makeig, S., Humphries, C., Lee, T.W., McKeown, M.J., Iraque, V., Sejnowski, T.J.: Removing electroencephalographic artifacts by blind source separation. Psychophysiology **27**, 163–178 (2000)
18. Delmore, A., Makeig, S.: EEGLAB: An open source toolbox for analysis of single-trial EEG dynamics including independent component analysis. J. Neurosci. Methods **134**(1), 9–21 (2004)
19. Bell, A.J., Sejnowski, T.J.: An information maximization approach to blind deconvolution. J. Neural Comput. **7**(6), 1129–1159 (1995)
20. Jung, Tzyy-Ping, Makeig, Scott, Humphries, Colin, et al.: Removing electroencephalographic artifacts by blind source separation. J. Psychophysiology **37**(2), 163–178 (2000)
21. Pfurtscheller, G., Neuper, C.: Motor imagery activates primary sensorimotor area in humans. J. Neurosc. Lett. **239**(2), 65–68 (1997)
22. Aydemir, O., Kayikcioglu, T.: Wavelet transform based classification o invasive brain computer interface data. Radioengineering **20**(1), 31–38 (2011)
23. Ting, W., Guozheng, Y., Banghua, Y., Hong, S.: EEG feature extraction based on wavelet packet decomposition for brain computer interface. Measurement **41**(6), 618–625 (2008)
24. Lakany, H., Conway, B.A.: Classification of wrist movements using EEG-based wavelets features. In: Proceedings of 27th IEEE EMBS Annual Conference, China, pp. 5404–5407 (2005)
25. Dong, S.C., Amato, V., Murino, V.: Wavelet-based processing of EEG data for brain-computer interfaces. In: Proceedings of IEEE Computer Society Conference, USA, pp. 74–74 (2005)
26. Yong, Y.P.A., Hurley, N.J., Silvestre, G.C.M.: Single trial EEG classification for brain-computer interface using wavelet decomposition. In: Procedings of EUSIPCO 2005, Eurasip, Antalya, Turkey, (2005)
27. Mallat, S.: A theory for multiresolution signal decomposition: the wavelet representation. IEEE Pattern Anal. Mach. Intell. **11**(7), 674–693 (1989)
28. Rumelhart, D.E., Durbin, R., Golden, R., Chauvin, Y.: Backpropagation: Theory, Architectures, and Applications, Hillsdale, Lawrence Erlbaum Association, New Jersey, pp.1–34 (1995)
29. Nicolaou, N., Nasuto, S.J.: Temporal independent component analysis for automatic artifact removal from EEG. In: Proceedings of 2nd International Conference on Medical Signal and Information Signal Processing, Sliema, Malta (2004)
30. Müller-Gerking, J., Pfurtscheller, G., Flyvbjerg, H.: Designing optimal spatial filters for single-trial EEG classification in a movement task. Clin. Neurophysiol. **110**(5), 787–798 (1999)
31. Ramoser, H., Müller-Gerking, J., Pfurtscheller, G.: Designing optimal spatial filters for single trial EEG during imagined movement. IEEE Trans. Rehab. Eng. **8**(4), (2000)
32. He, S.L., Xiaorong, G., Fusheng, Y., Shangkai, G.: Imagined hand movement identification based on spatio-temporak pattern recognition of EEG. In: IEEE Proceedings on EMBS, pp. 599–602 (2003)

Website

http://www.mathworks.com/

Negotiation-Based Patient Scheduling in Hospitals

Reengineering Message-Based Interactions with Services

Lars Braubach, Alexander Pokahr and Winfried Lamersdorf

Abstract Nowadays, hospitals in Germany and other European countries are faced with substantial economic challenges stemming from increased costs and increased demands inter alia resulting from a changing age pyramid. In this respect, patient scheduling is an interesting parameter that determines on the one hand the length of the patients stay in the hospital and the efficiency of the hospital resource allocation on the other hand. Due to the specific characteristics of hospitals such as unexpectedly arriving emergencies or unintended complications during treatments the optimization of patient scheduling is an extraordinary difficult task that cannot be easily solved using a typical centralized optimization algorithm. Thus, in this chapter a decentralized agent based approach is presented that represents patients as well as hospital resources as agents with individual goals that negotiate to find appropriate solutions. The approach has been extensively studied within the MedPAge project and also has been implemented using an agent platform. In this work it will be shown how the traditional message based implementation, which was difficult to construct and even more difficult to maintain, can be replaced with a service based design.

L. Braubach (✉) · A. Pokahr · W. Lamersdorf
Distributed Systems and Information Systems Group, Computer Science Department,
University of Hamburg, Vogt-Kölln-Str. 30, 22527 Hamburg, Germany
e-mail: braubach@informatik.uni-hamburg.de

A. Pokahr
e-mail: pokahr@informatik.uni-hamburg.de

W. Lamersdorf
e-mail: lamersdorf@informatik.uni-hamburg.de

B. Iantovics and R. Kountchev (eds.), *Advanced Intelligent Computational Technologies and Decision Support Systems*, Studies in Computational Intelligence 486, DOI: 10.1007/978-3-319-00467-9_10, © Springer International Publishing Switzerland 2014

1 Introduction

In hospitals, patient scheduling deals with the assignment of patients to the scarce and expensive hospital resources. Optimizing patient scheduling is able to reduce hospital costs due to a reduced stay time of patients within the hospital and an increased capacity utilization of hospital resources. Despite its general usefulness, efficient mechanisms for patient scheduling are difficult to devise due to some inherent characteristics of hospitals. Foremost, in hospital many uncertainties exist that make precise planning ahead for weeks or even just days very difficult if not impossible. These uncertainties e.g. result from unexpectedly arriving emergencies and treatment complications and may require far-reaching changes with respect to the patient scheduling plans. Another important difference with respect to other application domains with rather static workflows consists in the unpredictability of patient treatment processes within the hospital. Although approaches like clinical pathways [5] aim at standardizing the treatment steps of patients for certain kinds of diseases, patient cure remains a task that cannot be preplanned completely in advance. Thus patient scheduling has to accommodate these characteristics and provide a flexible and fast mechanism that is able to handle uncertainties at many levels.

Within the MedPAge project [7, 9, 14], patient scheduling is treated as a decentralized coordination problem between patients and hospital resources. Both types of stakeholders are modeled as autonomous decision makers with their own selfish goals that need to come to agreements in order to build up scheduling plans. In the following, the foundations of patient scheduling in hospitals will be described in Sect. 2. Afterwards, in Sect. 3, the MedPAge approach will be introduced by explaining its coordination mechanism, the general system architecture and its implementation based on multi-agent system technology. Implementation experiences showed that a message based realization of the complex coordination protocol is difficult and error prone. Hence, in Sect. 4 an alternative coordination description and implementation based on service interfaces is proposed and illustrated by dint of the MedPAge coordination mechanism. Section 5 summarizes and concludes the chapter.

2 Foundations

Hospitals typically consist of different rather autonomous units that are responsible for different medical tasks. Patients normally reside at wards and visit ancillary units according to prescribed treatments. Often, these treatments are prescribed by physicians as part of the ward round taking place in the morning of each day. The treatment requests are announced to the different ancillary wards, which decide on their own behalf at what time a patient will be served. For this purpose the ancillary unit typically employs a first come first served principle and calls the patients from the ward in the order the requests arrived. This scheme is very flexible and allows the ancillary wards to react timely to unexpectedly occurring emergencies by calling an emergency before normal requests and with respect to complications by simply delaying the next patient call as long as needed. Besides its flexibility the scheme also incurs drawbacks that derive from the local unit perspective used. This local view does not take into account global constraints e.g. the current allocations of other ancillary units or the overall treatment process of a patient.

Patient scheduling in hospitals aims at optimizing the temporal assignment of medical tasks for patients to scarce hospital resources with two objectives. On the one hand the patient's stay time and on the other hand the resource's idle times should be minimized. As both objectives are not necessarily congruent, a coordination approach can be used, in which both sides try to reach agreements by making compromises. In order to negotiate about the different time slots at scarce hospital resources the patients need to be able to quantify their interest in the resources. For this purpose the preferences of patients should be expressed in terms of the medical priority of the examinations or treatments for the patients. In this respect a concept for opportunity costs has been developed that is able to assess the difference of the patients health state with and without a potential treatment taking into account the duration of that treatment. In this way the potential decrease of a patients health state is used as a measure for the criticality of the treatment for that patient. As durations of treatments can vary in practice stochastic treatment durations have been taken into account. Details about the design of medical utility functions can be found in [7].

3 MedPAge Approach

A major objective of the MedPAge project consisted in finding an adequate decentralized coordination approach that on the one hand respects the decision autonomy of the units and on the other hand is able to generate anytime solutions that are at least near to pareto-optimal assignments. In order to come closer to this aim different coordination mechanisms [6, 8, 9] have been designed and benchmarked with respect to the current practice in hospitals. The benchmarking was based on a simulation system that used collected real world hospital data from a middle sized German hospital. The benchmarking revealed that an auction based scheme similar to contract-net performed best and was able to outperform the current hospital practice from 6 to 10 percent depending on the number of emergencies occurring (the more emergencies the lower the gain) [7]. Furthermore, the approach had been field tested in a hospital and was considered as a helpful decision support system by the staff [14]. In the following the details of the conceived coordination mechanism will be described (cf. [7]).

3.1 Coordination Protocol

In general, each hospital resource auctions off timeslots for treatment and examinations in near real-time, i.e. shortly before the ongoing treatment has finished and the resource expects to become available again. A time slot is assigned to the patient that placed the highest bid according to the health-state based utility functions introduced in the Sect. 3. The coordination protocol is designed to have four phases (cf. Fig. 1).

During subscription phase patients first have to find resources fitting to the treatments or examinations currently needed. Due to the fact that multiple medical units may offer the same treatments or examinations, a patient can subscribe at several units at the same time. Although the hospital infrastructure is considered as quite stable in comparison to the continuous arrival and dismissal of patients, it is assumed that from time to time changes can occur, e.g. a new medical unit is added or the operation of a unit is temporary closed. For this reason patients should

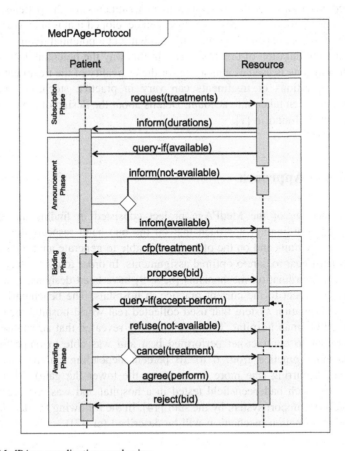

Fig. 1 MedPAge coordination mechanism

always perform a fresh search for adequate resources and afterwards subscribe at the corresponding resources. A resource answers to the patient with an estimate of the duration for the tasks requested. These durations are deduced from historical data and consist of mean and variance to cope with their stochastic nature.

In the announcement phase a resource initiates a new auction whenever the ongoing medical task is about to finish. In this way the flexibility of the current practice in hospitals is kept and possibly occurring disturbances caused by emergencies or complications can be addressed in the same way as before. The resource agent announces the auction by requesting which patient is currently willing to participate. This could be a subset of the subscribed patients as some of them could already have won an auction at another resource.

After having determined the participants, a call for proposal is sent to them with detailed information about the auctioneered treatment including the expected duration. Afterwards the patients are expected to send bids according to their utility functions, i.e. patients determine their bids according to the expected utility loss if they would loose the auction. This disutility is calculated by comparing the health state at the assumed finished time with and without the treatment.

The resource collects the incoming bids and sorts them according to the bid value. It will then start determining the winner in the awarding phase by notifying the first patient from the list. If the patient is still available it will agree and the resource will inform the other patients about the auction end. If not, the resource will move on with the next patient from its list and continue in the same way as before until a winner could be determined or no more patients are available. In this case the auction has failed and the resource could start over again with a new one.

3.2 System Implementation

The original MePAge system architecture is depicted in Fig. 2 and consists of three layers with different functionalities. The complexity of the layered model arises through different aspects. First, as already stated, the system has been built as simulation for benchmarking different coordination mechanisms and also as normal desktop system that could be used in field tests within a hospital. In this respect, it should be possible to reuse as much of the functional parts as possible when switching from simulation to operation. Second, MedPAge was part of a higher-level initiative called Agent.Hospital, in which several projects were integrated into a more complete hospital environment. Hence, the interoperability of the system was an important aspect solved by a shared hospital ontology.

Due to the different use cases, distinct user interfaces have been built for simulation and real world system testing. The simulation user interface allowed for performing simulation experiments and evaluating the statistical result whereas the user interface for operation was conceived as a decision support system (cf. Fig. 3). It can be seen that different views have been developed according to the functional roles involved within the hospital. From an administrative point of

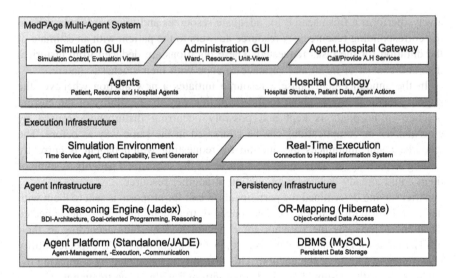

Fig. 2 System architecture

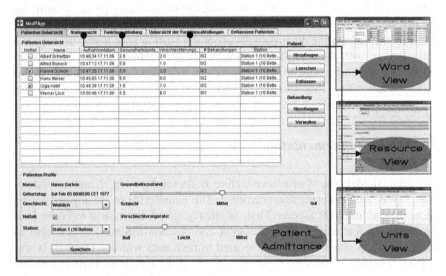

Fig. 3 System screenshot

view the system allows for entering new patient and resource master data. This was necessary as the system has not been directly connected to the hospital server infrastructure, which would have allowed reusing existing data. Within the ward view, the current health state according to the diagnosis can be coarsely given and required treatments and examinations can be added to the patients in correspondence to the medical prescriptions. The system automatically generates appointment proposals that become visible within the unit as well as in the ward view.

In the unit view the currently waiting patients are shown to the personnel and it allows for synchronizing the real hospital resource with the virtual one. For this purpose it can be entered to system when a medical task begins and ends or if a disruption exists. In the ward view the system announces treatment calls for patients according to the internally applied auction mechanism. The personnel is free to accept or reject the system proposals and send patients to the resources as they deem appropriate. These choices should be entered into the system in order to allow for correct future planning.

The core of the MedPAge system is the coordination mechanism that has been realized using a multi-agent negotiation approach. Hence, patient as well as resource agents have been modeled and implemented as representatives for their corresponding stakeholders. Details about the agent based design can be found in [4]. The coordination mechanism has been realized as message based interaction protocol in accordance to the sequence diagram shown in Fig. 1. As execution infrastructure the Jadex agent platform [11] has been used, which originally relied on the JADE agent platform [1]. Hospital data and scheduling information was stored in a relational database (MySQL), which had been interfaced with an object-relational mapper (Hibernate). In order to support the simulation as well as real time execution an additional execution infrastructure middle layer had been built.

3.3 Weaknesses and Challenges

The experiences gained by implementing the MedPAge system led to several important research questions that were found to require addressing them also on a general layer and not only in a project specific manner:

- Simulation and Operation of a system should be possible without modifying the functional core of the system. Rebuilding the system from a simulation prototype leads to new failures and much additional implementation overhead.
- Implementing coordination mechanisms using asynchronous message passing is a very tedious an error prone task. Even experienced developers struggle in foreseeing all possible protocol states leading to increased development efforts and delays in software production.

In order to address the first challenge we worked on the notion of simulation transparency [12] denoting the fact that a simulation and real system become indistinguishable from a programmers point of view (except its connection to a real or virtual environment). The underlying idea consists in introducing a clock abstraction within the infrastructure that is used for all timing purposes. Exchanging the type of clock (e.g. from a real time to an event driven mode) immediately changes the way time is interpreted in the system. Following this path made obsolete the execution infrastructure layer in case of the MedPAge system.

The second challenge requires even more fundamental changes in the way systems are designed and implemented. The proposed solution path consists in changing the communication means used in the system. On the one hand the original idea of MedPAge, namely having negotiating patients and resources should be kept, but on the other hand the communications should be simplified. To achieve this objective, a service oriented perspective is combined with the agent oriented view leading to the notion of active components described in the following section. Relying on active components the complex interaction scheme can be largely simplified by using services. It has been shown that conceptually that agents can be equipped with services without loosing their special characteristics like autonomy [3].

4 Service-Oriented Interaction Design

In this section, a redesign of the MedPAge system is sketched that takes into account the lessons learnt from its original implementation. Here, we focus on the redesign of the coordination mechanism according to the newly developed active components approach. Therefore, the basic concepts of the active components approach are shortly presented in the Sect. 4.1. Afterwards, a straight forward method is proposed how to map message-oriented negotiation protocols to appropriate service interfaces. The resulting service-oriented redesign of the MedPAge coordination protocol is put forward and the advantages of the approach are discussed.

4.1 Active Components Approach

The active components approach is a unification of several successful software development paradigms, incorporating ideas from object, components, services and agents [2, 10]. Therefore, it combines features such as the autonomous execution from agents and the explicit specification of provided and required interfaces from components (cf. Fig. 4). An important characteristic of the approach is that each active component may act as service provider and service user at the same time. Thus the interaction of active components is not limited to the client/ server model as used in, e.g., web services, but also naturally supports peer-to-peer communication as found in multi-agent systems. Regarding the architecture of an application, the required and provided service interfaces of active components allow making component dependencies explicit and easily manageable already at design time. Yet, by supporting dynamic service search and binding, the model allows for adaptive applications that automatically build and change (parts of) their structure at runtime in response to a dynamically changing environment (e.g. when service providers disappear or new services become available).

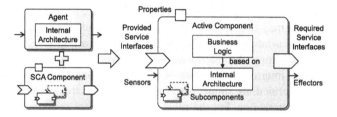

Fig. 4 Active component structure

The active components approach has many intentional similarities to the standardized and industry-supported service component architecture (SCA[1]). Yet, there are important differences due to the incorporation of agent ideas. In the active components approach, each active component is an autonomously executing entity (like an agent) that interacts with other (possibly remote) active components by calling methods of their provided service interfaces. For autonomy reasons and also for avoiding consistency issues due to concurrent access to a components internal state, active components follow a single-thread execution model per component, which means that external requests and internal actions are queued and executed sequentially (cf. [3]). To respect this execution model and also to support distribution transparency, i.e. no difference in programming for remote versus local service calls, all service operations should use asynchronous method signatures. Asynchronous methods can be realized using so called *futures* [13] as return values, i.e. objects that act as proxies for later results of ongoing operations. Once the asynchronously invoked operation finishes, the actual result will be made available in the future object.

In Jadex,[2] which is our Java-based reference implementation of the active components approach, different kinds of futures are provided that capture recurring interaction patterns in a natural way. In the following, three important types are introduced by dint of a simple, illustrative example of a wearable heart rate monitor connected wirelessly to a stationary display device. The heart monitor service allows retrieving the last observed heart rate and also creating an electrocardiogram (ECG) plot of the historic heart activity according to a specifiable period. In addition, the display can subscribe to the heart rate monitor for being sent periodic updates of the current heart rate. As shown in Fig. 5 (line 2), the *getHeartRate()* method of the monitoring service returns a simple future of type double. Because the service execution is done asynchronously, the rate value is not immediately available. Instead, the display component may add a listener to the future (line 3) for being informed, when the service execution is finished and the result can be obtained. For long running operations, the terminable future as used in the *generateECGPlot()* method (lines 4, 5) allows terminating an ongoing

---

[1] http://www.oasis-opencsa.org/

[2] http://jadex-agents.informatik.uni-hamburg.de

operation, while it is still in progress (cf. line 7). This kind of interaction is useful, when service calls may take a long time to complete and allow, e.g., the user to abort the requested printing of the ECG plot and choose a shorter period. Finally, for recurrent interactions, the intermediate future type allows the service implementer to publish intermediate results of potentially long-lasting operations. A special kind of intermediate future is the subscription future, which resembles a publish/subscribe interaction. For the *subscribeToHeartRates()* method (line 9, 10) you can see that the display is informed about each new heart rate value in the *intermediateResultAvailable()* method. To cancel such a subscription, also the *terminate()* method of the future can be used, i.e. every subscription future is also an instance of the terminable future type.

4.2 Modelling Interactions with Services

In message-based interactions, each communicative act is represented as a simple message. As inspired by speech-act theory, difference between those acts can be identified by using performatives such as *query* or *inform*. The allowed sequences of sent and received messages can be specified in interaction protocols, which can be specified, e.g., in AUML sequence diagrams. The foundation for intelligent physical agents (FIPA) has proposed a standardized set of performatives and also defined several commonly used interaction protocols like *request* and *contract-net*. Despite many attempts for providing tools or other development support, the current practice of implementing of message-based interactions still exhibits a large gap between design and the actual programming. Message-based interaction is not naturally supported by today's prevalent programming languages like Java and C#. This requires implementing many aspects of the interaction hidden in

```
01:  IHeartMonitorService heartservice;
02:  IFuture < Double> rate = heartservice.getHeartRate();
03:  rate.addResultListener(...);

04:  ITerminableFuture<ECGPlot> generateplot
05:   = heartservice.generateECGPlot(long period);
06:  ...
07:  generateplot.terminate();

08:  ISubscriptionIntermediateFuture<Double> heartrates
09:   = heartservice.subscribeToHeartRates();
10:  heartrates.addResultListener(new IntermediateResultListener() {
11:    public void intermediateResultAvailable(Decimal rate) {
12:      // update display value...
13:    }
14:  })
```

Fig. 5 Examples of future types

application code. As a result, implementation errors, such as type errors or invalid message sequences, can only be detected at runtime, if at all. This makes implementing message-based interactions a tedious and error-prone, time consuming task.

On the other hand, for the communicative acts in service interactions, different representations exist for, e.g., a service call, its return value or potential exceptions. The future patterns introduced above allow further types of communicative acts, such as termination of ongoing operations or publication of intermediate results. All these types of interaction are naturally represented by well-established object-oriented concepts like interfaces, method signatures and futures. Because these constructs are all first class entities of current object oriented programming languages, sophisticated tool support exists for statically checking the soundness of interaction implementations already at compile time (e.g. checking number and type of parameters, compatibility of return values or the types of futures supported for an interaction).

For the above reasons, we consider a service-based implementation of an interaction advantageous to a message-based one. Still the design of a service-based interaction is a non-trivial task, because in addition to deciding about sequences of communicative acts, also each act needs to be mapped to one of many fundamentally different interaction types, such as service calls, return values, etc. Therefore we propose an approach starting from a message-based interaction design and converting it to a service-based design for easier implementation. In this way the decisions about which communicative acts are required and how these are represented are straightened, thus simplifying the design process. In addition, the design of message-based interactions is well-understood and it seems reasonable to reuse the existing concepts, standards and tools from this area.

As a result from practical experiences in converting message-based designs to service-based ones, we identified the following recurrent steps:

1. *Identification of service provider*: In almost every case, the initiator of an interaction acts as client of a participant, i.e. the first communicative act of an interaction is best represented as a service call from the initiator to the corresponding participant.

2. *Mapping of subsequent communicative acts*: For each subsequent communicative acts it needs to be decided how it is represented according to the following options:

   a. Messages from initiator to participant can be represented as service calls or as termination request to previous service calls.
   b. Messages from participant to initiator can be represented as service result, intermediate service result or as failure (exception) of a previous service call.

3. *Evaluation of the resulting design:* The design should be evaluated according to the following design criteria

    a. Establishing high cohesion of methods in any interface and low coupling between different interfaces.

    b. Minimizing dependencies between methods (e.g. avoid that methods are only allowed to be called in a special ordering).

    c. Avoiding to introduce new communicative acts (e.g. when an act is mapped to a service call, but there is no message corresponding to the return value).

4. *Splitting Interfaces:* If no appropriate design can be found with a single service interface, the interaction needs to be split up into several interfaces. Therefore one or more messages in the interaction need to be chosen as start of a sub-interactions and the process needs to be repeated from step 1 for each sub-interaction.

Of course, in practice these steps should be considered as guidelines and not as a rigid process. For each concrete interaction, probably different designs need to be considered to identify advantages and limitations and to decide on the best option for implementation.

4.3 Redesign of the MedPAge Coordination Mechanism

The service-based redesign process has been applied to the MedPAge coordination mechanism as presented in Sect. 3.1. The resulting interfaces are shown in Fig. 6. The design has been obtained by following the process outlined in the Sect. 4.2. The interaction is initiated by a patient, which has some treatments without a corresponding reservation at a resource, yet. Therefore the resource is the participant of the interaction and should provide a service interface to the patient (lines 1-5). During the redesign it became obvious that some of the messages of the original interaction are obsolete, especially in the announcement phase. Therefore these messages were omitted during the redesign. In the original interaction protocol, the resource would periodically start auctions in which the registered patients could bid for an upcoming time slot. This naturally maps to a service call by the patient to register a required treatment and in response the resource would provide a subscription future to publish the start of each auction (lines 2, 3). In order to simplify the interaction and also to allow for dynamic changes in the resource operation, it was decided, that the stochastic treatment durations are supplied by the resource to the patient at the start of each auction. Therefore the *subscribeToAuctions()* method of the resource service corresponds to the initial *request(treatment)* message. It should be noted that the Jadex active components infrastructure automatically makes available the caller of a service to the service implementation and thus no separate parameter for identifying the patient that called the resource method is necessary. The subscription future returned by the method represents both the *inform(durations)* message as well as the *cfp(treatment)* message. To include information about the auctioneered time slot and the

```
01:  public interface IResourceService {
02:    public ISubscriptionIntermediateFuture<CFP>
03:      subscribeToAuctions(String treatmenttype);
04:    public ITerminableFuture<Boolean> bid(CFP cfp, double costs);
05:  }

06:  public interface IPatientService {
07:    public IFuture<Boolean> callPatient();
08:  }

09:  public class CFP {
10:    public String treatmenttype;
11:    public long startdate;
12:    public double duration;
13:    public double variance;
14:  }
```

Fig. 6 Redesigned MedPAge coordination mechanism

corresponding treatment durations, the *CFP* object type is introduced (lines 9-14) and used as result type of the subscription future.

When a new auction starts, the patient receives this CFP object calculates its opportunity costs and sends its bid to the resource. This is done using the *bid()* method (line 4) corresponding to the *propose(bid)* message. The result of the method can be interpreted as corresponding to the *query-if(accept-perform)* and *reject(bid)* message in case of success or failure. Yet, this would leave the three remaining messages *refuse(not-available)*, *cancel(treatment)* and *agree(perform)* to be realized by an isolated method in the resource interface. This would have violated design criteria 3b (minimizing dependencies), because the required ordering of the three resource interface methods would not have been obvious from their syntactic structure. Instead it was decided, that the majority of the awarding phase was better represented by a separate patient interface (lines 6-8). Here the resource can inform the winner of an auction using the *callPatient()* method corresponding to the *query-if(accept-perform)* message. The patient may answer with the boolean result value if it is available or if the resource should prefer another patient. The advantage of this design is that the call-patient functionality can also be used independently of the auction mechanism. E.g. a resource that wants to apply a different scheduling scheme may simply accept registrations in the *subscribeToAuctions()* method, but never start any auction and instead call patients directly as deemed appropriate.

4.4 Discussion

In the following, the specific improvements of the MedPAge coordination algorithm are discussed. In the original message-based MedPAge coordination protocol, twelve different messages between two roles have been used. Without the obsolete announcement phase still nine messages remain. The redesigned service-oriented coordination mechanism captures the same functionality, but only requires a total of three methods spread over two interfaces. Thus, the perceived complexity in the service-based interaction design is much lower than in the message-based design. Furthermore, the object-oriented service interfaces are conceptually much closer to the implementation technology, which even more reduces the complexity for the programmer. On a conceptual level, the new design keeps the agent-oriented view of patient and resource representatives, which negotiate for finding bilaterally appropriate schedules. Therefore, the autonomy of the different hospital units is preserved. Yet, the object-oriented design makes dependencies between the communication acts also syntactically clear. E.g., the *subscribeToAuctions()* method needs to be called before the *bid()* method, because the latter requires the *CFP* object supplied by the former method.

5 Conclusion

In this chapter patient scheduling has been identified as important area of improvement in hospitals. The currently employed technique for patient scheduling is a first-come first-served scheme that has the advantages of being highly flexible and respecting the autonomy of the hospital units. Within the MedPAge project a decentralized auction-based solution has been developed, which preserves these characteristics but additionally adds a global optimization perspective. The MedPAge system has been built as an agent-oriented solution. Based on the original system design and implementation some of its weaknesses and resulting challenges have been identified. The active components approach and in particular the service-oriented interaction design have been presented. They address the weakness of the complexity and implementation effort that is induced by message-based interaction designs. A service-oriented redesign of the MedPAge coordination mechanism has been developed and its advantages have been discussed.

During the course of the MedPAge project, the agent metaphor was found a highly suitable design approach for decision support systems in the area of hospital logistics. The agents can act as representatives of the involved hospital stakeholders, taking into account their respective goals and the typical local autonomy of existing hospital structures. One remaining obstacle for putting agent-based systems into productive use is still the gap between agent technology and established mainstream technologies, such as object-orientation and, e.g., web services. The service-based approach presented in this chapter supports an easier integration with legacy software, allowing, e.g., to use web services for provided or required service calls.

References

1. Bellifemine, F., Caire, G., Greenwood, D.: Developing Multi-Agent systems with JADE. Wiley, New York (2007)
2. Braubach, L., Pokahr, A.: Addressing challenges of distributed systems using active components. In: Brazier, F., Nieuwenhuis, K., Pavlin, G., Warnier, M., Badica C. (eds.) Intelligent Distributed Computing V—Proceedings of the 5th International Symposium on Intelligent Distributed Computing (IDC 2011), pp. 141–151. Springer (2011).
3. Braubach, L., Pokahr, A.: Method calls not considered harmful for agent interactions. Int. Trans. Syst. Sci. Appl. (ITSSA) 1/2(7), 51–69 (2011)
4. Braubach, L., Pokahr, A., Lamersdorf, W.: MedPAge: Rationale Agenten zur Patientensteuerung. Knstliche Intelligenz 2, 33–36 (2004)
5. Dept of Veteran's Affairs, Australia: Clinical pathway manual for geriatric community nursing. www.dva.gov.au/health/provider/provider.htm (2000)
6. Paulussen, T.: Agent-based patient scheduling in hospitals. Ph.D. thesis, Universitt Mannheim (2005)
7. Paulussen, T., Zller, A., Rothlauf, F., Heinzl, A., Braubach, L., Pokahr, A., Lamersdorf, W.: Agent-based patient scheduling in hospitals. In: Kirn, S., Herzog, O., Lockemann, P., Spaniol, O. (eds.) Multiagent Engineering—Theory and Applications in Enterprises, pp. 255–275. Springer, Berlin (2006)
8. Paulussen, T.O., Heinzl, A., Rothlauf, F.: Konzeption eines koordinationsmechanismus zur dezentralen ablaufplanung in medizinischen behandlungspfaden (medpaco). In: 5. Internationale Tagung Wirtschaftsinformatik 2001, pp. 867–880. Physica, Heidelberg (2001).
9. Paulussen, T.O., Zller, A., Heinzl, A., Pokahr, A., Braubach, L., Lamersdorf, W.: Dynamic patient scheduling in hospitals. In: Bichler, M., Holtmann, C., Kirn, S., Mller, J., Weinhardt, C. (eds.) Coordination and Agent Technology in Value Networks. GITO, Berlin (2004)
10. Pokahr, A., Braubach, L.: Active components: A software paradigm for distributed systems. In: Proceedings of the 2011 IEEE/WIC/ACM International Conference on Intelligent Agent Technology (IAT 2011). IEEE Computer Society (2011).
11. Pokahr, A., Braubach, L., Lamersdorf, W.: Jadex: A BDI reasoning engine. In: Bordini, R., Dastani, M., Dix, J., El Fallah Seghrouchni, A. (eds.) Multi-Agent Programming: Languages, Platforms and Applications, pp. 149–174. Springer, Berlin (2005)
12. Pokahr, A., Braubach, L., Sudeikat, J., Renz, W., Lamersdorf, W.: Simulation and implementation of logistics systems based on agent technology. In: Blecker, T., Kersten, W., Gertz, C. (eds.) Hamburg International Conference on Logistics (HICL'08): Logistics Networks and Nodes, pp. 291–308. Erich Schmidt Verlag (2008)
13. Sutter, H., Larus, J.: Software and the concurrency revolution. ACM Queue 3(7), 54–62 (2005)
14. Zller, A., Braubach, L., Pokahr, A., Rothlauf, F., Paulussen, T.O., Lamersdorf, W., Heinzl, A.: Evaluation of a multi-agent system for hospital patient scheduling. Int. Trans. Syst. Sci. Appl. (ITSSA) 1, 375–380 (2006)

A New Generation of Biomedical Equipment Based on FPGA. Arguments and Facts

Marius M. Balas

Abstract The chapter is aiming to broaden the bridge that covers the gap between the engineering and the biomedical science communities, by encouraging the developers and the users of biomedical equipment to apply at a large scale and to promote the Field-Programmable Gate Array technology. The chapter provides a brief recall of this technology and of its key advantages: high electrical performances (great complexity, high speed, low energy consumption, etc.), extremely short time-to-market, high reliability even in field conditions, flexibility, portability, standardization, etc. The positive FPGA experience, issued from the military and the aerospace domains, is beginning to spread into the biomedical and healthcare field, where the personnel should be aware and prepared for this substantial and presumably long term advance.

1 Introduction

A quite shocking title is drawing our attention when reading the second quarter 2009 issue of IEEE Circuits and Systems Magazine: "A Wake-Up Call for the Engineering and Biomedical Science Communities" [1].

The author of this chapter are trying to sensitize the readers on the gap that is drawn between the new achievements of electronics and information technology on one hand and the biomedical science community on the other hand. In order to fill this gap the authors envisaged a coherent action: the establishment of the Life Science Systems and Applications Technical Committee (LISSA) within the IEEE Circuits and Systems Society, supported by the USA National Institutes of Health (NIH). LISSA organizes several workshops each year. After each workshop a

M. M. Balas (✉)
Aurel Vlaicu University, Elena Dragoi, 310330 Arad, Romania
e-mail: marius@drbalas.ro

B. Iantovics and R. Kountchev (eds.), *Advanced Intelligent Computational Technologies and Decision Support Systems*, Studies in Computational Intelligence 486, DOI: 10.1007/978-3-319-00467-9_11, © Springer International Publishing Switzerland 2014

white paper is published by the IEEE Circuits and Systems Magazine in order to present the major challenges in various chosen theme areas.

Inspired and alerted by this message, we are now proposing an idea that could lead to a comprehensive modernization and improvement of the biomedical equipment: the large scale applying of the Field-Programmable Gate Array (FPGA) technology. The unification of the major part of the equipment used by a whole scientific-technical domain, by means of the FPGA, is already experienced with very positive results by the aerospace industry [2, 3] etc.

2 The FPGA Technology and its Place in the Electronic Industry

FPGA is an electronic technology that is essentially implementing very large scale custom digital circuits by means of the software controlled reconfiguration of a large integrated array, composed of identical Configurable Logic Blocks (CLB). A CLB cell may contain different types of Look-Up-Tables (LUT), flip-flops and elementary logic gates. The FPGA routing system ensures the reprogrammable CLBs configuration that defines each application, by a wide network of horizontal and vertical channels that may be interconnected in any possible way by transistorized interconnecting matrices. In other words, FPGA is a field reprogrammable Very Large Scale Integrated Circuit (VLSI), merging some leading hardware and software technologies. A FPGA generic structure is shown in Fig. 1 [4].

In the late 1980s the US Naval Surface Warfare Department initiated a research project that succeeded to build a computer out of 600,000 reprogrammable gates—the direct ancestor of the FPGA hardware. The FPGA software's origin is alike: although the first modern Hardware Description Language (HDL), Verilog, was introduced by Gateway Design Automation in 1985, the decisive leap occurred in 1987, when the US Department of Defense funded a research aiming to lead towards a unified language for description and simulation of the digital circuits, especially Application-Specific Integrated Circuits (ASICs). This approach generated Very High Speed Integrated Circuit Hardware Description Language (VHDL). Using VHDL all the military equipment, no matter its producers, could be treated in the same optimal manner (conception, design, testing, implementation, maintenance, upgrading, etc.). Verilog and VHDL were used basically to document and simulate circuit-designs initially described in another forms, such as schematic files or even analytic functions. After that, a lot of synthesis tools were developed in order to compile the HDL source files into manufacturable gate/transistor-level netlist descriptions. A key contribution to the FPGA development belongs to IEEE that marked the concept's evolution by conceiving the initial version of VHDL (the 1076–1987 standard), followed by many other subsequent related standards.

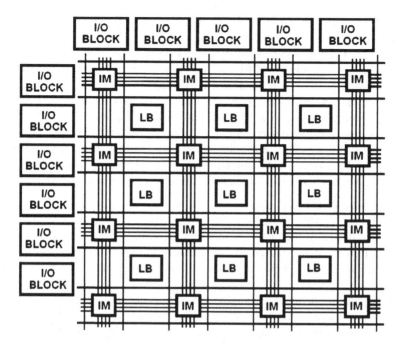

Fig. 1 FPGA generic structure, with logic blocks LB, wired channels, interconnecting matrices IM and input/output interface I/O

The company that imposed in 1985 the first commercial FPGA—XC2064 is Xilinx (co-founders Ross Freeman and Bernard Vonderschmitt). XC2064 gathered for the first time programmable gates and programmable interconnects between gates. Xilinx is the tenure leader of the FPGA market, strongly involved into the aerospace industry [2, 3]. A turning point in the public perception of FPGAs happened in 1997, when Adrian Thompson merged the genetic algorithm technology and FPGA to create the first sound recognition device. Other popular FPGA producers are: Altera, Actel, Lattice Semiconductor, Silicon Blue Technologies, Achronix and QuickLogic.

It is useful for us to relate FPGA with the other leading electronic and IT technologies, in order to point its pros and contras if used in biomedical equipment.

Gate array is a fundamental concept for digital integrated circuits. Since the Transistor Transistor Logic (TTL) times, the integrated circuits were realized starting from gate arrays—NAND gates for instance, simply by executing the wired connections that were configuring the desired electric schematics. Wirings were initially mask-programmed at factory, mainly by photolithography or other Read Only Memory (ROM) technologies. Given the initial gate array, the design of a new integrated circuit comprised essentially only the design of the photolithographic mask of the wiring system. Replacing this rigid technology with the flexible software controlled configuration of the today's FPGA followed the next steps, always in connection with the developments of the ROM memories:

- The Programmable Array Logic (PAL), sharing with Programmable ROM (PROM memories) the mask-programmable feature.
- The Field Programmable Array Logic (FPAL) issued providing PALs with appropriate programmers. FPALs are one time programmable: users can write their custom program into the device only once;
- The Complex Programmable Logic device (CPLD) were higher capacity PALs;
- In the respect of the parallelism with the memories technologies, FPGAs are field reprogrammable devices, similar to Erasable PROM (EEPROMS): users can repeatedly write and erase the stored data referring to the configuration. However, FPGAs outreached by far the simple field reprogrammable devices; their reconfiguration circuits became extremely complex and performing.

Besides the PAL-CPLD connection, FPGAs is organically linked with ASICs. The ASICs are VLSIs customized for a particular use. Customization occurs by the appropriate design of the metal interconnect mask, there are no reprogrammable features. This ROM like technology was developed in terms of sheer complexity (and hence functionality) increasing the number of integrated gates to over 100 millions.

ASICS are meant to be customized for large scale applications or standard products, cell phones for instance. Their design and development are laborious and expensive, but their electrical performances are rewarding: fast operation, low energy consumption, extremely high reliability, etc. The high end of the ASIC technology belongs to the full-custom ASIC design, which defines all the photolithographic layers of the device. The full-custom design minimizes the active area of the circuit and improves its functional parameters by minimal architectures and the ability to integrate analog components and pre-designed components, such as microprocessor cores.

However, the amount of applications that can use exactly the same circuit in huge number is limited, so usual ASICs are realized by gate array design. This manufacturing method uses unconnected wafers containing all the diffused layers, produced by full-custom design, in extremely large quantities. These wafers are stocked and wait to be customized only when needed, by the interconnection layout. Thus, the non-recurring engineering costs (research, development, design and test) are lower, as photolithographic masks are required only for the metal layers, and production cycles are much shorter. On the other hand this technique is absolutely rigid, the smallest design or implementation error compromises the whole batch of circuits.

This is why the first motivation of the PFGA developers was to create rapid prototyping ASICs, in order to avoid the wastage and to further reduce the time to market and the nonrecurring costs. FPGAs can implement any logical function that an ASIC could perform, so when developing an ASIC, the first mandatory stage is to realize its FPGA version. The FPGA prototypes can be easily reconfigured and tuned, until obtaining the desired performances and successfully completing all the testing specifications. Only the final versions of the FPGA netlists (all the connections of the circuits) will be automatically implemented into ASICs. As a matter of fact this technological itinerary holds only for large scale applications, very often the medium or small scale applications remain as FPGAs.

A key issue when comparing FPGA to other electronic technologies is the FPGA versus μC-DSP relationship. FPGA that is stemming from the digital integrated circuits achieved today a strong position into the embedded systems field. By contrast with a general-purpose computer, such as a PC, which is meant to be flexible and to meet a wide range of end-user needs, an embedded system is a computer system designed to do one or a few dedicated and/or real-time specific functions, very often in difficult environment and operation conditions. Each time when we need powerful algorithms, complex data and signal processing or intelligent decisions at the level of physical applications, embedded systems become inevitable. The embedded systems were developed by the microprocessor μP—microcontroller μC—digital signal processor DSP connection. A legitimate question for the biomedical equipment developers is why to replace the μP-μC oriented generation with a new FPGA oriented one.

Comparing the two solutions one can observe that computer/μC binary architectures are build around a central spine bone: the data/address/control bus (see Fig. 2). All the instructions of the program use one by one the bus, in order to accomplish the four steps that form the instruction cycle: fetch, decode, execute and writeback of the resulting data. Except the decoding, all the other steps demand complex bus traffic and maneuvers, which creates a fundamental limitation of the computer's speed.

The inconvenient of the bus centered architecture is that since only one instruction is executed at a time, the entire CPU must wait for that instruction to complete before proceeding to the next instruction. Caches memories, pipelines and different parallel architectures such as multi-core systems are addressing with more or less success this inconvenient. For the embedded systems frame a suitable solution represent the parallel bus that carries data words in parallel, on multiple wires. If we compare Figs. 1 and 2 we may appreciate FPGAs as fully distributed and parallel, totally opposed to the bus centered architectures. Comparing to the

Fig. 2 The generic computer/microcontroller architecture

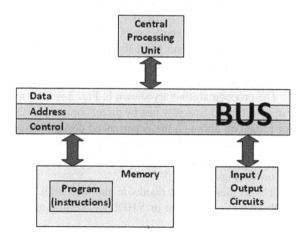

programmed operation of a μC, the FPGA presents a programmed architecture. The FPGAs are replacing the virtual computing of the logic functions that follow the synchronized succession of the instructions and all their steps, with a real wired circuit, able to operate continuously, with no complicated synchronization constraints. This is fundamentally accelerating the FPGA performance. The same algorithm may be performed hundred times faster by a FPGA circuit compared to usual bus oriented architecture devices.

However, the FPGA shift is paid by rather laborious programming, which implies a specific vision. A relevant synthesis of the digital computer versus FPGA dichotomy was provided by Klingman [6]: When gates were precious entities and tools 100 % proprietary, it made ultimate sense to arrange these limited gates into universally used objects, such as CPU registers, ALUs, instruction decoders (Arithmetic Logic Unit), and address decoders. You would then provide a set of instructions that linked and relinked these elements, so that, for example, two CPU register outputs could be connected (via buses) to an ALU input, then the ALU output connected to a destination register, and then the ALU input connected (via buses) to a specific memory address, and the ALU output connected to a different register, and so on and so on ...

When gates are no longer precious but are commodities, the fixed elements approach no longer makes as much sense. The monstrous development in languages, tools, I/O devices, standards, and so on will keep CPU development and implementation alive for decades, if not centuries, but the economics are now and trending more so in the FPGA direction.

3 FPGA Design Tools

The FPGA's intrinsic complexity, pointed by ED Klingman in 2010, is addressed in 2012 by design tools that become more and more friendly. The last edition of Xilinx ISE Design Suite for instance is justifying its name by gathering a bunch of previously stand alone software tools, covering all the operations demanded by the design and the implementation of a FPGA application. The main stand alone tools embedded into ISE are: ISim (behavioral simulation), PlanAhead (mapping and post-layout design analysis), iMPACT (creates PROM files) and ChipScope (post analyze of the design).

The procedural itinerary shown in Fig. 3 is fully automate, a simple application can be achieved basically just by direct commands (Generate Post-Place and Route for instance), although the experienced designers have full access to the software resources (Manually Place and Route for instance). The main stages of a FPGA application development following the previous itinerary are illustrated in the following Figs. 4–9 [7, 8].

It is to mention that thanks to the high HDL standardization the conversions HDL code ↔ schematic or VHDL ↔ Verilog are very fast and reliable.

Fig. 3 The Xilinx ISE
design suite procedural
itinerary

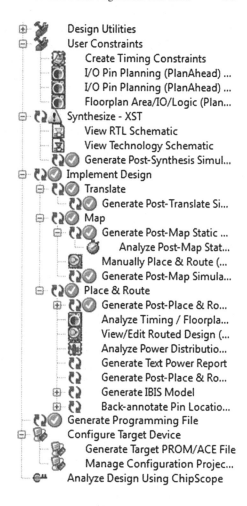

Design Utilities
User Constraints
    Create Timing Constraints
    I/O Pin Planning (PlanAhead) ...
    I/O Pin Planning (PlanAhead) ...
    Floorplan Area/IO/Logic (Plan...
Synthesize - XST
    View RTL Schematic
    View Technology Schematic
    Generate Post-Synthesis Simul...
Implement Design
    Translate
        Generate Post-Translate Si...
    Map
        Generate Post-Map Static ...
            Analyze Post-Map Stat...
        Manually Place & Route (...
        Generate Post-Map Simula...
    Place & Route
        Generate Post-Place & Ro...
        Analyze Timing / Floorpla...
        View/Edit Routed Design (...
        Analyze Power Distributio...
        Generate Text Power Report
        Generate Post-Place & Ro...
        Generate IBIS Model
        Back-annotate Pin Locatio...
    Generate Programming File
Configure Target Device
    Generate Target PROM/ACE File
    Manage Configuration Projec...
Analyze Design Using ChipScope

Another proved approach is the automate conversion from the well known C code to the low-level FPGA programming files, more precisely to HDL [8]. This conversion creates conditions for a massive transfer of knowledge (algorithms, procedures, etc.) from the C written applications camp towards FPGA, although critical and optimized FPGA applications need the direct intervention of the experienced designers.

4 Existing Healthcare and Biomedical FPGA Applications

The leading producers (Xilinx, Altera, Digilent, etc.) have reached a maturity level that enables FPGAs to deal with a great amount of applications, stemming from

(a)

```
15   LIBRARY ieee;
16   USE ieee.std_logic_1164.ALL;
17   USE ieee.numeric_std.ALL;
18   LIBRARY UNISIM;
19   USE UNISIM.Vcomponents.ALL;
20   ENTITY RSlatch_RSlatch_sch_tb IS
21   END RSlatch_RSlatch_sch_tb;
22   ARCHITECTURE behavioral OF RSlatch_RSlatch_sch_tb IS
23
24       COMPONENT RSlatch
25       PORT( XLXN_1    :   OUT   STD_LOGIC;
26             XLXN_2    :   OUT   STD_LOGIC;
27             XLXN_3    :   IN STD_LOGIC;
28             XLXN_4    :   IN STD_LOGIC);
29       END COMPONENT;
30
31       SIGNAL XLXN_1   :   STD_LOGIC;
32       SIGNAL XLXN_2   :   STD_LOGIC;
33       SIGNAL XLXN_3   :   STD_LOGIC;
34       SIGNAL XLXN_4   :   STD_LOGIC;
35
36   BEGIN
37
38       UUT: RSlatch PORT MAP(
39           XLXN_1 => XLXN_1,
40           XLXN_2 => XLXN_2,
41           XLXN_3 => XLXN_3,
42           XLXN_4 => XLXN_4
43           );
```

(b)

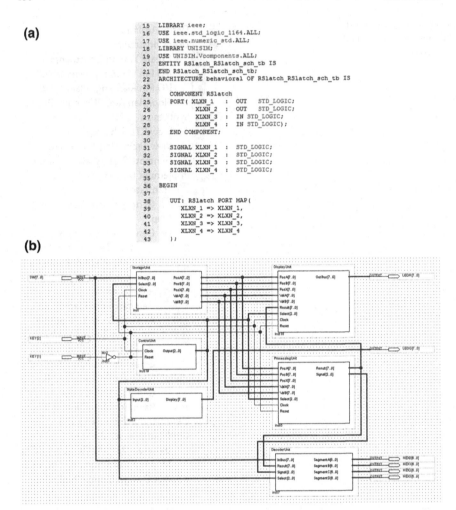

Fig. 4 The synthesis of the HDL code/schematic source of the application

virtually any possible domain, including the Artificial Intelligence. The biomedical field was taken seriously into consideration by all producers. In 2007 Altera team promoted FPGAs as mathematical parallel coprocessors in different models of medical equipment, boosting the speed hundred times, lowering up to 90 % the power consumption comparing to conventional coprocessors, with better utilization of the resources in many types of algorithms [9].

Image processing is one of the best suited applications for FPGAs, due to the parallel architectures, the logic density and the generous memory capacity. Besides driving any type of high resolution display devices, FPGAs are capable to boost the performances of all the rapid imaging algorithms: image construction, enhancement and restoration, image analysis and pattern recognition, imagine and data

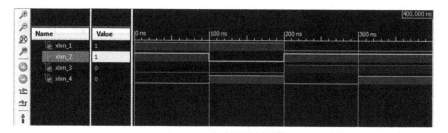

Fig. 5 The behavioral simulation (ISim)

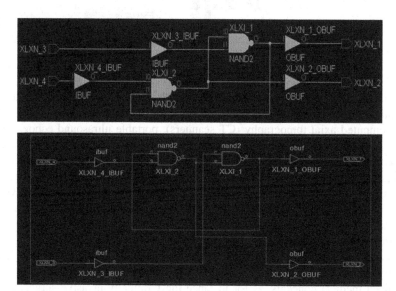

Fig. 6 The RTL schematic (register transfer level) and the reviewed schematic

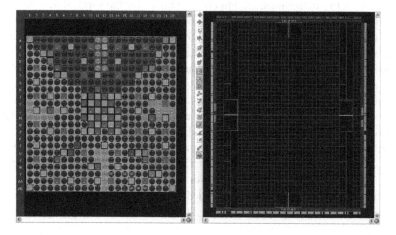

Fig. 7 The mapping of the design (assigning I/O pins) and the place and route

Fig. 8 Generating the netlist and installing the bitstream file into the FPGA's ROM

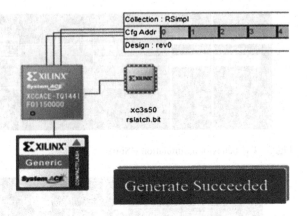

compression, color space conversion, etc. Virtually any significant medical imaging technique can be successfully coped by FPGAs: X-rays, magnetic resonance imaging, computed axial tomography (CT scanners), portable ultrasound echographic devices, 3D-imaging, surgical microscopes, optical measuring and analyze instruments, telemedicine systems, etc. [10, 11], 4D imaging-based therapies [12], etc.

Besides imaging, a wide range of clinical applications are mentioned in the FPGA oriented literature: electro surgery equipment [11], heart assisting devices [12], robotic surgery, portable patient monitoring systems, drug delivery systems, automatic external defibrillators (AED) [13], bio-sensors using SmartFusion programmable analog devices [14], etc.

Due to their flexibility and fast upgradeability and to their extremely short time-to-market, the FPGAs are now called to support the development of the most advanced medical concepts, as for instance the minimally-invasive surgery platforms (MIS), which means significantly less trauma and faster recovery for patients. A MIS platform combines the above mentioned techniques (robotics, endoscopy, 3D visualizations, etc.) in order to minimize the surgical actions and to maximize their precision and sharpness. Such a platform, called da Vinci Robotic Surgical System, is now produced by Intuitive Surgical. The platform is composed by an ergonomically designed console where the surgeon sits while operating, a patient-side cart where the patient lays during surgery, four interactive robotic arms, a high-definition endoscopic 3D vision system, and a set of surgical instruments, adapted to the seven degrees of freedom robotic wrists. So far this platform has enabled methodologies for cardiac and general surgery, urology, pediatric, gynecology, colorectal, and head and neck surgeries [11].

The portable ultra-low-power FPGAs enable the implementation of a variety of devices that minimize the power consumption down to 2 μW when the system is not in active use, an important issue for systems such as automated external defibrillators, which may be left unattended for weeks or months between tests or for the portable insulin pumps [13].

Fig. 9 Popular FPGA boards. **a** Altera DE1. **b** Xilinx Spartan 6

5 Conclusions

The FPGA industry reaches now the potential that enables it to deeply revolutionize all our technologies. The healthcare and biomedical equipment industry is facing a strategic challenge: the replacement of conventional electronic equipment characterized by bus centered architectures with FPGA/ASIC based systems, bringing substantial advantages: low costs, flexibility, interoperability, reliability, speed, etc.

Acknowledgments This work was supported by the Bilateral Cooperation Research Project between Bulgaria-Romania (2010–2012) entitled "Electronic Health Records for the Next Generation Medical Decision Support in Romanian and Bulgarian National Healthcare Systems", *NextGenElectroMedSupport*.

References

1. Chen, J., Wong, S., Chang, J., Chung, P.C., Li, H., Koc, U.V., Prior, F.W., Newcomb, R.: A wake-up call for the engineering and biomedical science communities. IEEE Circuits Syst Mag, second quarter, 69–77 (2009)
2. Xilinx and LynuxWorks Demonstrate Avionics Application Solution at the Military and Aerospace Forum and Avionics USA Conference 2009. Lynux Works, http://www.lynux works.com/corporate/press/2009/xilinx.php
3. Gerngross, J.: Intellectual property offers new choices for avionics design engineers as MIL-STD-1553 and FPGA technologies converge. Military Aerosp. Electron. **14**(6) (2003). http://www.militaryaerospace.com/index/display/article-display/178482/articles/military-aerospace-electronics/
4. Rodriguez-Andina, J.J., Moure, M.J., Valdes, M.D.: Features, design tools, and application domains of FPGAs. IEEE Trans. Ind. Electron. **54**(4), 1810–1823 (2007)
5. Hennesy, J.L., Patterson, D.: Computer architecture: A quantitative approach. Morgan Kaufmann (2006)
6. Klingman, E.: FPGA programming step by step. Electronic Engineering Times. http://www.eetimes.com/design/embedded/4006429/FPGA-programming-step-by-step
7. Xilinx: ISE In-Depth Tutorial. 18. Jan. (2012)
8. Balas, M.M., Sajgo, B.A., Belean, P.: On the FPGA implementation of the fuzzy-interpolative systems. In: 5th International Symposium on Computational Intelligence and Intelligent Informatics ISCIII Floriana, Malta, pp. 139–142, 15–17 Sept (2011)
9. Strickland, M.: Medical applications look towards FPGA-based high-performance computing. Hearst Electronic Products, 18 Dec (2007)
10. Altera Corp. white papers: Medical Imaging Implementations using FPGAs. July (2010)
11. Khan, K.: FPGAs help drive innovation in complex medical systems. Med. Electron. Des. April (2012)
12. Microsemi: Size, Reliability and Security. Jan (2011). www.acaltechnology.com/
13. Actel: Incredible Shrinking Medical Devices. Oct (2008)
14. Microsemi: Intelligent Mixed Signal FPGAs in Portable Medical Devices. Application Brief AC 242, Dec (2010)

A Survey of Electronic Fetal Monitoring: A Computational Perspective

Kenneth Revett and Barna Iantovics

Abstract Electronic Fetal Monitoring (EFM) records fetal heart rate in order to assess fetal well being in labor. Since its suggestion in clinical practise by de Kergeradee in the nineteenth century, it has been adopted as standard medical practise in many delivery scenarios across the globe. The extent of its use has augmented from its original purpose and is now used to not only reduce prenatal mortality, but also neonatal encephalopathy and cerebral palsy. One of the difficulties with EFM is interpreting the data, which is especially difficult if it is acquired in a continuous fashion. A grading system has been developed (utilised for developing a guideline for Clinical Practise Algorithm) which consists of grading fetal heart rate (FHR) into fairly rough categories (three). These categories are defined by values associated with a set of four features. The values for this set of features are potentially influenced by the particular collection equipment and/or operating conditions. These factors, in conjunction with a stressful condition such as a complicated delivery scenario may render rapid and unequivocal reporting of the neonatal status sometimes difficult. This chapter examines the development of automated approaches to classifying FHR into one of three clinically defined categories. The ultimate goal is to produce a reliable automated system that can be deployed in real-time within a clinical setting and can therefore be considered as an adjunctive tool that will provide continuous on-line assistance to medical staff.

Keywords Biomedical datasets · Cardiotocogram · Decision support systems · Electronic fetal monitoring · Reducts · Rough sets

K. Revett (✉)
Faculty of Informatics and Computer Science, The British University in Egypt,
Suez Desert Road, 11837 Cairo, Egypt
e-mail: ken.revett@bue.edu.eg

B. Iantovics
Faculty of Sciences and Letters, Petru Maior University, Nicolae Iorga 1,
540088 Targu Mures, Romania
e-mail: ibarna@science.upm.ro

B. Iantovics and R. Kountchev (eds.), *Advanced Intelligent Computational Technologies and Decision Support Systems*, Studies in Computational Intelligence 486, DOI: 10.1007/978-3-319-00467-9_12, © Springer International Publishing Switzerland 2014

1 Introduction

The first modern day deployment of fetal monitoring began in the early nineteenth century by de Kergeradee, whom suggested that listening to the baby's heartbeat might be clinically useful [1]. He proposed that it could be used to diagnose fetal life and multiple pregnancies, and wondered whether it would be possible to assess fetal compromise from variations in the fetal heart rate. Today, monitoring the fetal heart during labour, by one method or another, appears to have become a routine part of care during labour, although access to such care varies across the world.

There are two basic types of fetal monitoring currently deployed routinely; intermittent auscultations (IA) and cardiotocography (CTG). With IA, the fetus is examined using a fetal stethoscope (a Pinard) at some frequency: typically every 15 min at the start of labour and more frequently (every 5 min) during delivery, though these figures are still debatable [2, 3]. This was the predominant form of fetal monitoring prior to the development of the CTG approach in the late part of the twentieth century. The CTG provides a continuous written recording of the fetal heart rate (FHR) and uterus activity (uterine contractions–UC). Note that the CTG can be recorded externally, using a Doppler type of device which is attached to a belt placed around the mother's abdomen (external CTG). This tends to restrict the mother's movement and prevents the use of a calming bath prior to delivery. During the actual delivery stage, after the amniotic sack has ruptured (either naturally or induced), a CTG clip can be placed on a presenting part (typically the head) and the FHR recorded (typically no UC data is acquired at this time). This form of CTG is termed an internal CTG.

Given this methodological approach to fetal monitoring, what information acquired from EFM should be utilised to render a decision as to the health status of the fetus? Essentially, there are four features which have been defined within the Guideline for Clinical practise, which are described in Table 1 [3]. The features are collected using standard medical equipment (termed a CTG) as described above, and are converted into a global feature which categorises the health status of the fetus (the first column in Table 1). This global feature is then utilised to generate a discrete category of the fetal status over each time point during the delivery monitoring period. There are three mutually exclusive categories which are depicted in Table 2.

The categories defined in Table 2 form a set of guidelines used by medical practioners to assess the health status of the fetus on a general level during delivery. The question is whether these guidelines are sufficiently robust to provide an accurate assessment of the fetal status. One approach to addressing this problem is to evaluate large datasets that contain feature information collected across a large sample of subjects. This chapter first examines typical feature sets that have been made publically available and secondly, it presents a few randomly selected case studies where these publically available datasets have been used for computational analysis. The chapter ends with a brief discussion of the impact computational studies have produced in terms of informing the current set of guidelines.

Table 1 Summary of the fetal heart rate (FHR) features utilised to characterize the current health status of the fetus (see [3] for details)

Feature	Baseline value (bpm)	Variability (bpm)	Decelerations	Accelerations
Reassuring	110–160	≥5	None	Present
Non-reassuring	100–109 161–180	<5 for >40 to <90 min	Early deceleration variable deceleration single prolonged deceleration up to 3 min	The absence of accelerations with an otherwise normal CTG are of uncertain significance
Abnormal	<100 >180	<5 for ≥90 min	Atypical variable decelerations Late decelerations Single prolonged Single prolonged deceleration > 3 min	

Table 2 Nomenclature and definitions of the fetal health status (N, S, & P) indicated by the data tabulated in Table 2. (see [3] for details)

Category	Definition
Normal	A CTG where all four features fall into the reassuring category
Suspicious	A CTG whose features fall into one of the non-reassuring categories and the remainder of the features are reassuring
Pathological	A CTG whose features fall into two or more non-reassuring categories or one or more abnormal categories

2 Publically Available CTG Datasets

The UCI data repository is a reputable portal for freely available datasets collected from a wide range of domains [4]. In particular, there is a CTG dataset at the UCI data archive, which consists of the CTG records of 2,126 patients acquired up to and including actual delivery (acquired on the day of delivery). It should be noted that this dataset has been reported most frequently as the data source for publications available over the internet [5–9]. For each patient record, a classification was assigned based on pre and post-partum evaluation by a qualified and licensed medical practitioner (obstetrician). The feature set consists of nominal and real-valued features which focus on the four principle features of CTG (baseline, accelerations, decelerations, and variability). In addition, the FHR and derivative features such as histograms were computed off-line, yielding a total of 21 features in the dataset. There were no missing values, all data was acquired under identical conditions and involved the same trained medical staff, minimising inter-subject variation. There were three decision classes (assigned by medical professionals): Normal, Suspect, and Pathological, with an object distribution of 78, 14, and 8 % respectively (1,655, 295, and 175 for each case respectively).

From a computational perspective, the dataset is typical in the biomedical domain. There are a fairly large number of features (21), and though there are no missing values, the three categories are not equally represented in terms of the cardinality of each decision class. For instance, the 'Normal' class contained 78 % of the data, significantly biasing the data in that direction. Despite this unequal distribution of decision class exemplars, most reported studies indicate a very high classification rate, along with relatively high values for the positive and negative predictive value [10–12]. These results obtained from a variety of difference approaches probably reflects the direct mapping of the features collected to the basis for rendering the decision. That is, the physician will base his decision on the definitions in Tables 1 and 2, and the features extracted are exactly those features. Judging by the consistently high classification accuracy of this dataset, acquired from diverse approaches such as neural networks, decision trees, support vector machines, rough sets, the relationship between the features and the decision class is supported by the data. This is a very encouraging result which could be exploited for the deployment of a real-time EFM system suitable in a clinical setting.

One approach to developing such a real-time monitoring system (and given the low sampling frequency, the 'real-time' adjective is not terribly critical) is to deploy a rule-based approach. There are several interesting papers which provide a relatively flat rule base from which one can comprehend and utilise quite efficiently in an expert system based approach. Our lab has applied a rough sets based approach to classifying the UCI based CTG dataset. The results of this study yield 100 % classification accuracy, with a PPV of 1.0 and a NPV of 1.0 as well [13]. The feature set of 21 values was reduced to 13, which were roughly evenly spread across all four major features (see Fig. 1). Other machine learning based systems have yielded similar values for classification accuracies over 99 % in many cases, but typically yield much lower values for PPV and NPV (see [10–12] for examples). The advantage of a rule-based approach, as opposed to a more recalcitrant approach such as Neural Networks is that these approaches (rule-based) reveal the underlying features and their values in the classifier per se. Note that rough sets (the approach utilised in [13]) generates a rule-based system which typically yields very high values for classification accuracy, PPV, and PNV (see [14] for a tutorial based discussion of rough sets). There is no need to explore the weight space or perform parameter sweeping in order to determine which features are important in

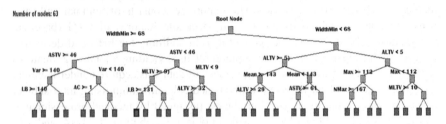

Fig. 1 The resulting decision tree generated from a rough sets approach to CTG data classification (see [13])

the resulting classification scheme. As an example of the output of a rule based system, a decision tree based classification result is depicted in Fig. 1.

3 Conclusions

That the ability of a variety of machine learning tools can extract useful and clinically relevant information from the large pool of available EFM data is without question. This type of data is extremely useful to both the machine learning community as well as the relevant domain (the medical) community. Rarely is there such as direct mapping of feature space onto decision classes, yielding such accurate and clear cut decisions. With the ability to map feature values to the major categories (Normal, Suspicious, or Pathological), the ultimate question to be addressed is the actual medical significance of the CTG in the context of patient care.

First and foremost, the CTG measures the fetal heart rate, which is a direct measure of fetal circulation. The FHR is categorised roughly into 3 disjoint cases: Normal, Suspect, or Pathological (N, S, or P respectively). These cases are based on measurements of a 4 valued feature set composed of: baseline value, short term variability, accelerations and decelerations. Further, another potential class is based on the response of the fetus to external stimuli such as contractions and fetal movements. The baseline feature is a record of the beats per minute, which tends to decrease during the last trimester up to deliver. The short term variability (STV) reflects the ongoing balance between accelerations and decelerations (via sympa/ parasympathetic activity controlled centrally) of the heart rate. Accelerations and decelerations occur naturally, when the fetus moves, the uterus contracts or the head moves down into the pelvic region. In addition, the administrations of drugs to the mother may induce changes in accelerations/decelerations. Based on these features which are recorded during CTG, subjects are classified as either: Normal, suspect, or pathological. The fundamental physiological variable that underlies these features is the oxygen delivery level to the fetus. If the umbilical cord is temporarily occluded, oxygen levels to the fetus are decreased (hypoxia) and the fetus will undergo hypercapnia (increase CO_2 levels), which in turn cause respiratory acidosis. Furthermore, reduced oxygenation will induce metabolic acidosis, both of which induce a redistribution of blood flow to the fetus to compensate for these altered physiological states. These changes are normal—even fetal movement may cause fluctuations in the hemodynamics. What is important to monitor is the magnitude and duration of these measurable responses. Prolonged bradycardia (reduced heart rate below 100 bpm) is a very serious condition which may require immediate medical intervention. The magnitude, frequency, and duration of these physiological responses are used to categorise the status of the fetus.

That the CTG can be utilised in a real-time environment in an automated fashion is supported by the encouraging classification accuracy obtained by a variety of systems (with an emphasis here on rule-based systems). Training a rule based system based on existing literature results is the first step in this process. These results must be verified on unseen data in order to assess the generalisability of the resulting rule set. The results from a variety of such studies indicate high accuracies, PPV, and PNV values. These results together support the hypothesis that automated CTG based EFM systems are clinically useful. With a trained system in place, applying the rule-set (or other suitably trained machine learning algorithm) is typically extremely fast with respect to the phenomenon under investigation (especially true for the frequency deployed in EFM). So, from a computational perspective, EFM, measured via CTG, is a solved problem, in terms of being able to classify the fetus into one of the categories depicted in Table 2 (N, S, & P). The question remains whether or not this technology can be deployed at a lower level and/or integrated into a more comprehensive framework such as deployment throughout the 3rd trimester. To the first point, whether the system can detect secondary outcomes such as: it can detect the underlying causes of changes which place the fetus into one category or another? Integrating changes of blood oxygenation, CO_2 levels, blood pH, etc. could be integrated into the monitoring system, informed by the CTG results. The results could be the development of a more biologically comprehensive model of the fetus, instead of reporting fiduciary marks at specific frequencies. As for the second issue, computerized antenatal care, the system could be deployed to assess secondary outcomes such as gestational age at birth, and neonatal seizures would be extremely helpful [Cochrane Study paper]. We are undertaking this work in our lab—but requires additional data be placed in a publically available repository to complete this work. For this to happen, computational biologists require the cooperation and efforts of the medical community.

Acknowledgments The authors would like to acknowledge the source of the dataset: the UCI KDD data repository, into which this dataset was deposited by faculty members at the University of Porto, Portugal.

The research of Barna Iantovics was supported by the project "Transnational Network for Integrated Management of Postdoctoral Research in Communicating Sciences." Institutional building (postdoctoral school) and fellowships program (CommScie)-POSDRU/89/1.5/S/63663, Financed under the Sectorial Operational Programme Human Resources Development 2007–2013.

References

1. Alfirevic, Z, Devane, D., Gyte, GML.: Continuous cardiotocography (CTG) as a form of electronic fetal monitoring (EFM) for fetal assessment during labour (Review) 1, Copyright © 2007 The Cochrane Collaboration, Wiley (2007)
2. Ito, T., Maeda, K., Takahashi, H., Nagata, N., Nakajima, K., Terakawa, N.: Differentiation between physiologic and pathologic sinusoidal FHR pattern by fetal actocardiogram. J. Perinat. Med. **22**(1), 39–43 (1994)
3. American College of Obstetricians and Gynecologists Task force on Neonatal Encephalopathy and Cerebral Palsy. Neonatal Encephalopathy and Cerebral Palsy: Defining the Pathogenesis and Pathophysiology, Jan (2003)

4. UCI: http://kdd.ics.uci.edu/
5. Tsuzaki, T., Sekijima, A., Morishita, K., Takeuchi, Y., Mizuta, M., Minagawa, Y., Nakajima, K., Maeda, K.: Survey on the perinatal variables and the incidence of cerebral palsy for 12 years before and after the application of the fetal monitoring system. Nippon Sanka Fujinka Gakkai Zasshi, **42**, 99–105 (1990)
6. Ayres-de-Campos, D., Bernardes, J., Garrido, A., De Sa, J.P.M., Pereira-Leite, L.: SisPorto 2.0: a program for automated analysis of cardiotocograms. J. Matern. Fetal Med. **9**, 311–318 (2000)
7. Georgoulas, G., Gavrilis, D., Tsoulos, I.G., Stylios, C., Bernardes, J., Groumpos, P.P.: Novel approach for fetal heart rate classification introducing grammatical evolution. Biomed. Signal Process. Control **2**, 69–79 (2007)
8. Warrick, P., Hamilton, E., Macieszczak, M.: Neural network based detection of fetal heart rate patterns. IEEE Trans. Biomed. Eng. **57**(4), 771–779 (2010)
9. Fontenla-Romero, O., Guijarro-Berdinas, B., Alonso-Betanzos, A.: Symbolic neural and neuro-fuzzy approaches to pattern recognition in cardiotocograms. Adv. Comput. Intell. Learn. Int. Ser. Intell. Technol. **18**, 489–500 (2002)
10. Ulbricht, C., Dorffner, G., Lee, A.: Neural networks for recognizing patterns in cardiotocograms. Artif. Intell. Med. **12**, 271–284 (1998)
11. Georgoulas, G., Stylios, C., Groumpos, P.P.: Integrated approach for classification of cardiotocograms based on independent component analysis and neural networks. In: Proceedings of 11th IEEE Mediterranean conference on Control and Automation, 2003, Rodos, Greece, 18–20 June (2003)
12. Jezewski, M., Wrobel, J., Labaj, P., Leski, J., Henzel, N., Horoba, K., Jezewski, J.: Some practical remarks on neural networks approach to fetal cardiotocograms classification. In: Proceedings of IEEE Engineering in Medicine and Biology Society, pp. 5170–5173 (2007)
13. Revett, K.: A rough sets based approach to analysing CTG datasets. Int. J. Bioinform. Healthc. (in press)
14. Komorowski J., Pawlak Z., Polkowski L., Skowron A.: Theory, knowledge engineering and problem solving. Dordrecht: Kluwer, 1991. Rough sets: A Tutorial, In: Pal, S.K., Skowron, A. (eds.) Rough Fuzzy Hybridization: A New Trend in Decision Making, Springer, Singapore, pp. 3–98, (1999)

Quantifying Anticipatory Characteristics. The AnticipationScope™ and the AnticipatoryProfile™

Mihai Nadin

Abstract Anticipation has frequently been acknowledged, but mainly on account of qualitative observations. To quantify the expression of anticipation is a challenge in two ways: (1) Anticipation is unique in its expression; (2) given the non-deterministic nature of anticipatory processes, to describe quantitatively how they take place is to describe not only successful anticipations, but also failed anticipations. The AnticipationScope is an original data acquisition and data processing environment. The Anticipatory Profile is the aggregate expression of anticipation as a realization in the possibility space. A subsystem of the AnticipationScope could be a predictive machine that monitors the performance of deterministic processes.

Keywords Anticipation · Diagnostic · Possibility · Prediction · Process

1 Introduction

The year is 1590. Hans and Zacharias Jansen make public a description of what will eventually lead to the future light microscope. It was not a cure for malaria, dysentery, or African trypanosomias; but without the microscope, progress in treating such and other diseases would have taken much longer. The microscope allows for "mapping" the reality invisible to the naked eye. It helps in the discovery of "new territory". Under the microscope's lens, the "world" becomes "larger." Understanding the "larger" world becomes a goal for a variety of sciences. Bacteria and cells are identified. It soon becomes clear that micro-

M. Nadin (✉)
Institute for Research in Anticipatory Systems, University of Texas
at Dallas, Richardson, TX, USA
e-mail: nadin@utdallas.edu
URL: www.nadin.ws

B. Iantovics and R. Kountchev (eds.), *Advanced Intelligent Computational Technologies and Decision Support Systems*, Studies in Computational Intelligence 486, DOI: 10.1007/978-3-319-00467-9_13, © Springer International Publishing Switzerland 2014

organisms constitute the vast majority of the living realm. It also becomes clear that sight—the most natural of human senses—implies awareness of resolution. Since the retina is made of cells, it is possible to distinguish between two (or more) microentities only if they are at least a distance of $150 \cdot 10^{-6}$ m; otherwise, the eye cannot distinguish between them. In order to make the "invisible" (what the naked eye cannot see given its cellular make-up) visible, a lens was used in order to obtain a larger image of what the eyes could not distinguish. It took 220 years before achromatic lenses (that Giovanni Battista Amici used in his microscope) made possible the realization that cells (which Robert Hooke called by this name in 1665) exist in all organisms. Today this is common knowledge; at that time it was revolutionary.

The microscope, used in physics as well as in biology, helps to describe the matter that makes up everything that is (The sophisticated electronic microscope uses electrons instead of light.). It does not help to describe behavior at the macrolevel of reality. It only assists humans in figuring out what things are made from and how these components behave. And this is exactly what distinguishes the *AnticipationScope*™ from the enlargement machinery deployed in order to get a better look at lower-scale aspects of reality, or to "see" far away (telescopes were only a beginning).

In a day and age of advanced molecular biology and genetic focus, the AnticipationScope can be understood as a "microscope" for the very broad composite spectrum of conditions affecting the adaptive system:Parkinson's disease, autism, schizophrenia, obsessive-compulsive disorder, post-traumatic stress disorder, Alzheimer's. It does not visualize viruses or bacteria, but behavior. By generalization, it is a scope for human performance. The list of what can be perceived via the AnticipationScope is open. Brain imaging—the in brain scope a variety of implementations and processing techniques—is an attempt at mapping the brain. It has helped in understanding brain functionality, neurological disorders, and the efficiency of pharmacological treatments.

However, it does not allow for a closer look at unfolding human action as an integrated function across multiple systems, including sensory perception, cognition, memory, motor control, and affect. It may well be that the AnticipationScope is a digital processing platform corollary to brain imaging; or at least that the knowledge gained through the AnticipationScope can be correlated with that gained through brain imagery, or through molecular biology methods.

2 What Do We Map?

The operative concept upon which the entire project is grounded is *anticipation*. Since the concept is not mainstream in current science, the following example serves as an explanation. Anticipation, as a life sciences knowledge domain, spans a broad range of disciplines: cognitive science, bioinformatics, brain science, biophysics, physiology, psychology, and cybernetics, among others. The concept

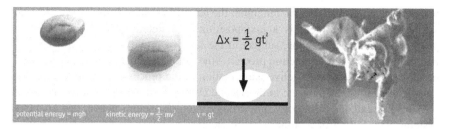

Fig. 1 Anticipation. *Left*: The falling of a stone. *Right*: The falling of a cat

challenges the reductionist-deterministic understanding of the living as a machine, subject only to physics, and the associated techniques and methods for "repairing" it. Given the concept's novelty, I shall explain anticipation using an example (see Figure 1)

The state of a *physical* (non-living) entity—the falling stone—is determined by its past state and current state (weight, position in space). The state of a living entity—the falling cat—is determined by its past and present, *but also by a possible future state* (safe landing). The falling of a cat involves what are known as adaptive processes [16]. The falling of a stone exemplifies a deterministic phenomenon. It can be described to a high degree of precision: the same stone, falling from the same position will fall the same way regardless of the context (day/night, no one looking/many observers, etc.).

$$x(t) = f(x(t-1), x(t)) \tag{1}$$

In the realm of physics, the experiment is reproducible. No matter how many times the stone falls, no learning is involved. A cat, while always subject to the laws of physics—the cat is embodied in matter, therefore gravity applies to its falling—exemplifies a non-deterministic phenomenon. Non-deterministic means that *the outcome cannot be predicted*. The cat might land safely, or it might get hurt. Even under the same experimental conditions, a cat will never fall the same way twice (not to mention that the cat that fell once is different from the cat that fell again—it aged, it learned). Adaptive characteristics change with learning (practice, to be more precise). Every experiment—assuming for argument's sake that someone might subject a cat to repeated falling (ethical considerations suspended)—returns a different value. Adaptive processes are involved. The cat's fall is associated with learning. Adaptivity changes with experience. As the context changes (falling during day or night, the number and type of observers, landing on grass or on stones, etc.), the outcome (how the cat lands) changes. Both a deterministic component and a non-deterministic component (the possible future state, i.e., where and how to land safely) affect the current state.

$$x(t) = f(x(t-1), x(t), x(t+1)) \tag{2}$$

In respect to the deterministic aspect: If the cat falls a short distance, its ability to influence the outcome is diminished. If it falls from a very high distance, the body's acceleration affects the outcome: the cat will get hurt despite trying to land safely. In summary: there is a juxtaposition of a deterministic action-reaction process-the falling of a stone, subject to the laws of gravity-and a non-deterministic, anticipation-action-reaction process—the falling of a cat is subject to gravity, but also subject to its own active participation in the process. In the first case, the knowledge domain is physics; in the second case, the knowledge domain involves physics, but extends to biology, physiology, motoric, neuroscience, and adaptive systems. The falling of a stone is representative of physical processes (a limited number of parameters relatively independent of each other). The falling of a cat is representative of complex processes: high number of variables, and usually reciprocally dependent variables. The mapping of repetitive processes-such as the falling of the stone-is fundamentally different from the mapping of unique complex processes. Let us now examine the conceptual underpinning of the anticipatory behavior example.

2.1 Conceptual Aspects

In respect to the history of interest (accidental or systematic) in anticipatory manifestations, one statement can be made: experimental evidence has been reported since the early beginnings of science. Theoretical considerations, however, have been more focused on particular forms of anticipation [9–11]. The earliest known attempts to define a specific field of inquiry [14] remain more of documentary relevance. It was ascertained very early that anticipation is a characteristic of the living. In this context, the focus on human movement is not accidental. Motor control remains an example of the many attempts to understand anticipatory aspects (such as those of the falling cat, mentioned above), even if anticipation as such is not named. In this respect, it is worth bringing up the so-called "Bernstein Heritage" [8]). Nikolai A. Bernstein's work (as contradictory as it is in its many stages) is a reference impossible to ignore (Fig. 2).

Since our particular focus is on quantifying anticipatory behavior, his kymocyclograph [2] deserves at least mention. It allowed him to take note of the uniqueness of human motion, replacing the mechanistic view with a dynamic view. The device is an original construction with many ingenious solutions to the problems posed by recording human motion. It is a camera conceived to allow

Fig. 2 *Kymocyclograph* conceived by Bernstein (1928) and hammering worker. *Left*: Kymocyclograph, *Right*: hammering worker

measurement of human movement. The focus was on "studying the *what* of movement (trajectories)" and "the *how* ('analysis of underlying mechanisms')" [4]. Of course, Bernstein's contributions are related to quite a number of other individuals (Popova, Leontief, Luria, Vygotsky, Gel'fand, Tsetlin, etc.).

It is worth mentioning that before Bernstein, Georges Marinescu [1], using a cinematographic camera, focused his studies (starting 1899) on gait disorders, locomotor ataxia, and other aspects of motoric neurological disorders. He was among the very first to recognize that film images could provide insight regarding the nature of changes in human movement (This reference does not necessarily relate directly to the subject of anticipation as such.). As already indicated, a theoretic perspective will eventually be advanced to those many researchers of human movement who already noticed anticipatory features in the way people move or execute operations using tools.

2.2 Anticipation and Change

Anticipation is an expression of change, i.e., of dynamics. It underlies evolution. Change always means variation over time. In its dynamics, the physical is subject to the constant rhythm associated with the natural clock (of the revolving Earth). The living is subject to many different clocks: the heartbeat has a rhythm different from saccadic movement; neuron firing in the brain takes place at yet another rhythm; and so does metabolism. The real-time clock that measures intervals in the change of the physical (day and night, seasons, etc.) and the faster-than-real-time clock of cognitive activity—we can imagine things well ahead of making them— are not independent of each other. A faster-than-real-time scanning of the environment informs the cat's falling in real time. A model of the future guides the action in the present. This is not unlike our own falling, when skiing downhill at high speed or when we accidentally trip. In the absence of anticipation, we would get hurt. Reaction is too slow to prevent this. The information process through which modeling the future takes place defines the anticipation involved. When this information process is affected by reduced perception, limited motor control, deteriorated reflexes, or short attention span, falling can become dangerous. Older people break bones when falling not only because their bones are getting frail (change in the physical characteristics), but also because they no longer "know" how to fall. In such cases, the anticipation involved is deficient.

3 Anticipation and Aging

It is in the spirit of this broad-stroke explanation of anticipation that a project entitled *Seneludens* (from *senescere*, getting old, and *ludere*, playing) was initiated by the antÉ - Institute for Research in Anticipatory Systems at the University of

Texas at Dallas. The gist of this project [11] is to compensate for aging-related losses in motor control, sensation, and cognition. The goal is to stimulate brain plasticity through targeted behavioral training in rich learning environments, such as specially conceived, individualized computer games (part of the so-called "serious games" development). These are customizable perceptual, cognitive, and motoric activities, with a social component that will strengthen the neuromodu-latory systems involved in learning. Improved operational capabilities translate into extended independence and better life quality for the aging. For this purpose, the Institute has formed research alliances with, among others, professionals involved in the medical support of the aging. From the many topics brought up in conversations with the medical community (e.g., University of Texas-South-western, a world renowned medical school, and Presbyterian Hospital, both in Dallas, Texas), one informs the goal of the endeavor: "How do you quantify anticipation? Isn't it, after all, reaction time?" (reflex time). It is not enough to assume that it is not.

Newton's action-reaction law of physics explains how the cat reacts on landing, but not how it anticipates the fall so that it will not result in pain or damage. The dimension of anticipation might be difficult to deconfound from reaction; but that is the challenge. And here lie the rewards for those who are engaged in addressing health from the perspective of anticipation. If we can quantify anticipation, we can pinpoint the many factors involved in the debilitating conditions affecting the aging (and not only the aging). The scientific hypothesis we are pursuing is: Spectrum conditions (e.g., Parkinson's, dementia, autism, etc.) are not the result of diminished reaction; rather all are expressions of reduced or skewed anticipation. Losses in sensorial acuity and discrimination, in motor control; diminished and deviating cognitive performance; affected memory functions—they all contribute to the loss of anticipation.

From all we know, anticipation is a holistic, i.e., integrated, expression of each individual's characteristics across multiple systems. Of course, the goal is to prove this statement, and based on the proof, to address, from the perspective of the unity between mind and body, the possibility to re-establish such unity.

3.1 Specifications

All of the above makes up the background against which the AnticipationScope introduced in the first lines of this text is defined: to conceive, design, specify, test, and implement an integrated information acquisition, processing, interpretation, and clinical evaluation unit. The purpose of measurement is to produce a record of parameters describing human action in progress. This is a mapping from processes that result in motoric activity to an aggregate representation of the action as meaningful information. The outcome of the AnticipationScope, within which an individual is tested, is his or her Anticipatory Profile™.

By way of explanation, let us consider current research in the genetic, or molecular, make-up of each human being. Indeed, medicine and molecular biology contributed proof that symptoms are important in dealing with disease; but molecular differences among individuals might be more important. We are past the one-drug-fits-all practice of medicine. The goal is to provide "personalized medicine" in which treatment is tailored not only to the illness, but also to the genetic or metabolic profile. Truth be told, this is an ambitious project. So is the Anticipatory Profile, but at the level of defining action characteristics of each person. The assumption that everyone runs the same way, jumps in an identical manner, hammers a nail in a standardized fashion, etc. is evidently misleading. If we could all run and jump the same way, champions would not be the exception. Their Anticipatory Profile is different from that of anyone else just trying to keep in shape. In Bernstein's time, the Soviet regime wanted scientists to study how to make every worker more productive (Machines do exactly that.). In our time, we want to understand how differences in anticipatory characteristics could lead to improved performance of artists, athletes, pilots, and of everyone enjoying a certain activity (tennis, skiing, hiking, swimming, golf, etc.).

Variations in the Anticipatory Profile are indicative of the adaptive capabilities of the individual. When we break down in our actions, it is useful to see whether this is due to an accidental situation or a new condition. Accordingly, the AnticipationScope could potentially identify the earliest onset of conditions that today are diagnosed as they eventually become symptomatic, usually years later. Indeed, as we shall see later on, Parkinson's is diagnosed six years after onset-only when it becomes symptomatic. By that time the process is irreversible.

This impedes the ability of the medical community to effectively assist those affected and eventually to cure them. Moreover, since we miss the inception of the condition affecting adaptive capabilities, we still do not know how they come about. In addition, the AnticipationScope can serve as a platform for evaluating progress in treatment. Data from the AnticipationScope can facilitate the type of inferences that the medical community seeks when addressing spectrum conditions. Regarding the functions of the AnticipationScope, we shall return to them once a more detailed description of what it is and how it works is presented.

3.2 The Design

Given the fact that anticipation is always expressed in action, the Anticipation-Scope combines the highest possible resolution capture of movement correlated with the capture of associated physiological and cognitive data. This sounds a bit complicated (not unlike the cMicroscope, announced by Changuei Yang and his team at the California Institute of Technology: a microscope on a chip with microfluidics). To explain as intuitively as possible what the AnticipationScope is, let us present another example: "... *when a man stands motionless upon his feet, if he extends his arm in front of his chest, he must move backwards a natural weight equal to that both natural and accidental which he moves towards the front,"* (Leonardo da Vinci 1498). His observation is simple: a physical entity modeled after a human being (in wood, stone, metal etc.) would not maintain its balance if the arms attached to this body would be raised (because the center of gravity changes). Figure 3 makes the point clear:

Leonardo da Vinci made this observation, prompted by his long-term study of the motoric aspects of human behavior. Five hundred years later, biologists and biophysicists addressing postural adjustment [6] proved that the compensation that da Vinci noticed—the muscles from the gluteus to the soleus tighten as a person raises his arm—slightly precedes the beginning of the arm's motion. In short, the compensation occurred in anticipation of the action. The AnticipationScope quantifies the process by capturing the motion, not on film (as did Marinescu and Bernstein) or video, but in the mathematical description corresponding to motion capture technology. Sensors, such as EMG (electromyography), goniometry, accelerometry, blood pressure, and EEG (electroencephalography) capture the various preparatory processes, as well as the reactive components of the balancing act (see Figure 4).

By design, the AnticipationScope is supposed to deliver an aggregate map of the human being recorded in action. Since anticipation is always expressed in action, such a map is indicative of the holistic nature of human actions, regardless of whether they are small tasks or elaborate compound endeavors. Behind the aggregated map is the understanding that in the past medicine focused on reductions. For instance, the cardiovascular profile (cf. Figure 5) is a mapping from parameters characteristic of blood circulation. Even today, preliminary to a physician's examination, a nurse or an assistant will, on a routine basis, check blood pressure and heart rate. These measurements are performed even before a dental check-up or an eye examination. Additionally, there are other partial mappings that a physician will consider: temperature, stress and hormonal profiles; a motor dynamics profile, if you visit an orthopedist's office (Fig. 5).

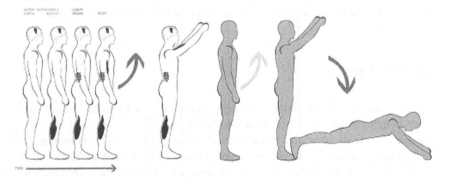

Fig. 3 *Dynamics of motion in the living and in the physical object (What does change in the center of gravity entail?). Left:* Dynamics of motion in the living. *Right:* Dynamics of motion in the physical

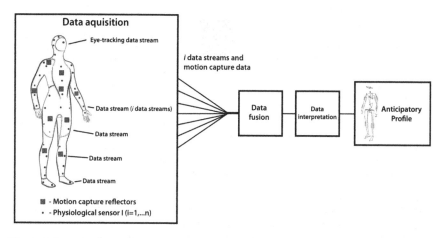

Fig. 4 Information processing model

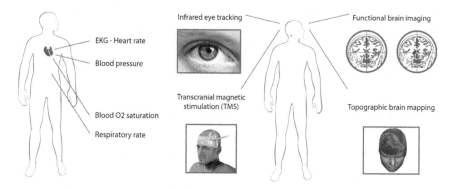

Fig. 5 *The cardiovascular profile* versus *The neural profile*. *Left*: The cardiovascular profile is a particular mapping focused on the heart and blood flow—a. *Right*: The neural profile

The measurement of body temperature (or of specific body parts), of galvanic skin response, or of muscle activity is indicative of physiological processes relevant to the diagnostic procedure. In recent years, given progress in imaging technology, the neural profile (Figure 5) is now possible. It integrates parameters significant to visual perception, to the performance of particular brain components (either under external stimulation or during experiments designed to engage such components), or to the brain as a whole. Evidently, the outcome of such measurements and evaluations is different in nature than determining blood pressure or body temperature. Each is already a composite expression of quite a number of processes involved in facilitating human action (Figure 6).

The Anticipatory Profile could be seen as an integrated map—although in its more comprehensive structure, it goes well beyond what each partial map provides in terms of meaningful information. The diagram Fig. 6 might suggest a simple additive procedure. In reality, the Anticipatory Profile is the expression of

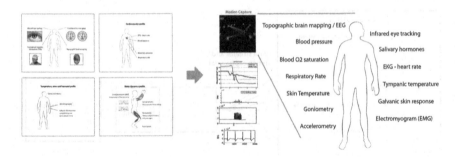

Fig. 6 Four profiles: aggregation of a cognitive and physiological map

integration and differentiation. Getting sweaty palms as we become aware that we have just been through a dangerous situation prompts more than galvanic response, even more than an accelerated heartbeat. All these partial descriptions—that is, time series of data associated with integrated physiological, neurological, and cognitive processes—contribute to a broader image of the individual. An identity comes to expression, quite different from person to person. The various aspects of data types and data integration will be discussed after examination of the architecture of the entity defined as the AnticipationScope.

3.3 Architecture of the Data Acquisition Platform

The research leading to an integrated mobile AnticipationScope starts with the double task of:

a. understanding the nature of subjecting the human being to measurement;
b. integrating unrelated technologies with the aim of producing a comprehensive representation of how the anticipation is expressed in action.

Each of the two tasks deserves further elaboration. Assuming that sensors could be devised to measure everything—given the dynamic condition of the living, this means an infinity of parameters—the problem to approach would be that of significance. In other words, how to reduce the infinite set of parameters to those that carry significant information about the anticipatory condition of an individual. The science necessary for this is inter- and cross-disciplinary. Therefore, the research team consisted of physicians, biologists, neuroscientists, and computer scientists. A large number of graduate students from the associated disciplines and from the game design program joined in the effort (some also served as subjects) (Fig. 7).

In respect to the architecture, the goal was to conceive a heterogenous data acquisition platform associated with the appropriate data processing and data interpretation facility. It was supposed to be an environment in which subjects carry out ordinary tasks: sitting on a chair, climbing stairs, throwing or catching a ball, walking, and running. Some tasks were to be performed on cue; others within

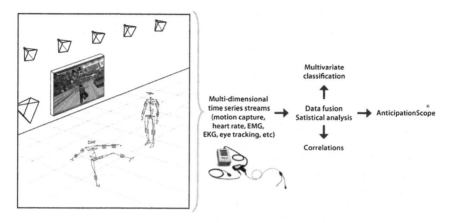

Fig. 7 Architecture of the integrated AnticipationScope

a "natural" sequence: climb stairs, reach a platform where sitting was available, "climbing down." Since to measure is to disturb the measured (Heisenberg's uncertainty principle of 1927), our focus was on minimizing the disturbance. This is possible because sensors of minimal weight are attached to the subject's clothing, and measurement does not affect the person subject to it or the parameters we want to evaluate. Examination of the architecture allows a simple observation: we are dealing with multi-dimensional time series streams. As a consequence, in order to make sense of the huge amount of data generated in real time, we need to provide data fusion procedures. Statistical analysis is the underlying mathematical foundation for producing multivariate classifications and for discovering correlations. In order to explain the technical challenge of the enterprise, let us shortly mention that the problems to be addressed correspond to (a) a great variety of data types; and (b) the need to find a common denominator for different time scales.

3.4 Data Types and Time Scale

A variety of sensors (for evaluating muscle activity, skin conductance, heart rate, respiration patterns, etc.) affords data characteristic of the measured parameters. The simple fact that an EMG sensor returns microvolts, while a skin conductance sensor returns micro-siemens, and the heart rate measured in beats per minute, is indicative of the nature of data acquisitions and the variety of measurement units. In computer science jargon, we deal with integers, Booleans, floating-point numbers, even alpha-numeric strings. Since we also evaluate color (of skin, eyes, etc.), we deal with a three-byte system (denoting red, green, and blue). Moreover, data types are associated with allowable operations (In some cases, as in color identification, addition and subtraction are permissible, but multiplication is not.).

Moreover, composite types, derived from more than one primitive type, are quite often expected. Arrays, records, objects, and more composite types are part of the data processing structure. The intention is to transcend the variety of concrete data types harvested in the AnticipationScope and eventually develop generic programming. We are frequently challenged by the need to calibrate data from one type of sensor (such as the motion capture sensors, 120 frames per second) with data from another type (EMG, for instance, at a much lower rate) before any integration can be carried out. Multiple data sources, distributed all over the body, also pose problems in respect to noise and wireless integration. Finally, the volume of real-time data generated within a session is such that storage, processing, self-correction procedures, and database management exclude the possibility of using readily available software.

As an example of how data integration is carried out let us consider only the aggregation of EMG and Motion capture data (Fig. 8):

Each EMG electrode measures the electric activity from the associated muscles. We follow a traditional measure to extract the feature of the EMG using the Integral of Absolute Value (IAV).

We calculate IAV separately for individual channels. Each channel corresponds to one EMG sensor. Let x_i be the ith sample of an EMG signal/data and N be the window size for computing the feature components. In a stream of EMG signal, let IAV_j be the *Integral of Absolute Value* of jth window of EMG, which is calculated as:

$$IAV_j = \frac{1}{N} \cdot \sum_{i=(j-1)\cdot N+1}^{j\cdot N} |x_i| \qquad (3)$$

With the global positional information for all segments, it becomes difficult to analyze the motions performed at different locations and in different directions. Thus, we do the local transformation of positional data for each body segment by shifting the global origin to the pelvis segment, since it is the root of all body segments. As a result, positional data of each segment with respect to global origin is transformed to positional data with respect to pelvis segment or joint as origin. The advantage of this local transformation is that the human motions were independent of the location of the participant in the room (which is helpful in some

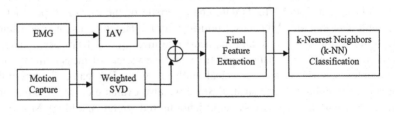

Fig. 8 Existing architecture for integrating motion capture and EMG data streams for quantifying the anticipatory characteristics of humans

tasks where the location of the participant in the room or the direction he/she is facing poses a problem).

An appropriate mapping function is required to map 3D motion joint matrices into 3D feature points in the feature space. In our implementation, we used the linearly optimal dimensionality reduction technique singular value decomposition (SVD) for this purpose. For any $l \times 3$ joint matrix A and window size N, the SVD for the jth window is given as follows:

$$A^j_{N \times 3} = U^j_{N \times N} \bullet S^j_{N \times 3} \bullet V^j_{3 \times 3} \tag{4}$$

S^j is a diagonal matrix and its diagonal elements are called singular values. For combined feature extraction techniques for EMG and motion capture data, we use the sliding window approach to extract the features from motion matrix data and EMG sensor data. To get a final feature vector corresponding to a window of a motion, we combine these two sets of features and map it as a point in multi-dimensional feature space, which is a combination of EMG and motion capture feature space. We do the fuzzy clustering using fuzzy c-means (FCM) on these mapped points to generate the degree of membership with every cluster for each point. Due to the non-stationary property of the EMG signal, fuzzy clustering has an advantage over traditional clustering techniques. For a given motion, range of highest degree of membership for each cluster among all the divided windows of a motion becomes the final feature vector for the given motion. The separability of these feature vectors among different motions depends on the fuzzy clustering. This extraction technique projects the effects, of both motion capture and EMG in a single feature vector for the corresponding motion (Fig. 9).

After the extraction of the feature vectors, similarity searching technique can be used to find the nearest neighbors and to do a classification for the motions. Our experiments show that classification rate is mostly between 80–90 %, which is understandable due to uncertainty of biomedical data and their proneness to noise. Some other unwanted environment effects such as signal drift, change in electrode characteristics, and signal interference, may affect the data. Other bio-effects, such as participant training, fatigue, nervousness, etc., can influence the effectiveness of

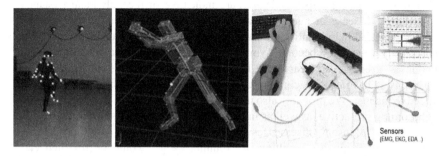

Fig. 9 *A session in the AnticipationScope, Left*: Subject of Experiment; *Middle*: Mapping of data *Right*: Integration of sensors

the biomedical activity. We also analyze the k-NN feature space classifier to check that, among the retrieved k-nearest neighbors, how many are a similar match or how many are in the same group of query. Since we consider the raw signal, the average percentage of correct matches among k-nearest neighbors is about 90 %.

4 The Unity Between Action and Perception

The basic premise of the entire design of the AnticipationScope is derived from the knowledge gained in recent years through single-neuron recordings [13] on the unity between action and perception. Brain activity is specific to the task embodied in the goal, not to the particular effectors. When an object is grasped, neurons are activated regardless of the hand (left or right), or of the toes (left or right foot) that might be used, or even if the mouth is used. *Purpose drives the action.* Anticipation is always purpose related. Gallese in [5] brings this observation to a clear point: *"The object representation is integrated with the action simulation."* This broad understanding of the unity of perception, activation processes, control mechanisms, and the motoric will guide the realization of the AnticipationScope as a space of interaction between the subject of inquiry and objects associated with actions.

Measurement within the AnticipationScope is goal related since anticipations are always in respect to the outcome of an action. Anticipation is not prediction, i.e., not a probabilistic process, but rather one driven by the evaluation of possibilities. The congenial perspective of the vast amounts of data collected (a variety of interactively selectable time correlated streams) is that of structural measurement process. The biologically inspired Evolving Transformation System (ETS) is the mathematical foundation for the structural measurement process [7]. ETS is a dynamic record of the interaction of elements involved in the functioning of a system, better yet in its self-organization. Each time we are involved in an action, learning takes place. The result is what we call experience [12].

As stated above, the output of the AnticipationScope is the individual Anticipatory Profile, which carries information about synchronization processes. Anticipation is an individualized expression, the "fingerprint" of human action. Variations in the Anticipatory Profile are indicative of the individual's adaptive capabilities. Disease or aging can affect the values. Accordingly, the AnticipationScope could help identify the earliest onset of conditions that today are diagnosed only when they become symptomatic—usually years later; in the case of Parkinson's disease, six years later. Delayed diagnosis (even of autism, despite its early onset) has negative consequences for the ability to assist in a timely and effective manner. We probably also miss important information that might guide us in becoming pro-active, as well as in finding a cure in some cases. To return to the example of Parkinson's disease: festination (loss of the center of gravity adaptive performance, eventually resulting in "running steps") could be revealed early through the AnticipationScope. Using data gathered from one sensor at a time—as practiced in describing, incompletely, ataxia, hemiparesis, dyskinesia, etc.—is a reductionist approach, useful but incomplete. Opposed to this is the integration of data from multiple sensors and from the motion capture that describes movements. This is the only way to capture the integrated nature of anticipation. Thus we gain a holistic understanding of the affected human being, in addition to specialized knowledge. The human being is a relational entity—while each component is relevant, the outcome is not reducible to the function of one or another part of our body, but rather to their interrelationship, how they synchronize.

5 Suggestive Directions

In using imaging and measurement, medicine acts upon representations. Data describing the body in action is a good representation for defining how well integration over many systems (neuronal, motoric, sensory, affective, etc.) takes place. In the spectrum condition, one or several systems either cease performing, or their performance no longer makes integration possible. Examining how the decline in anticipation results in maladaptive conditions is a life-science specific epistemological task. It challenges current scientific models. To make the point clear, here are some suggestive aspects, presented tentatively as information to be eventually tested in the AnticipationScope:

(1) Parkinson's Disease (PD)

    a. Early onset detection of akinesia, tremor, rigidity, gait and posture abnormalities. PD patients show impaired anticipatory postural adjustments due to deviated center of gravity realization, resulting in a delay in step initiation, or in "running steps";

    b. PD tremor is different from other forms of tremors (e.g., cerebellum tremors, essential tremors). This could be distinguished early with integrated motion capture and synchronized biosensors;

    c. PD gait: the typical pattern can be datamined by analyzing integrated motion capture and sensor data;

    d. Even minute PD-associated changes in the control system can be identified;

    e. Pre-pulse inhibition, also known as acoustic startle, could be an early indicator for diagnosing PD. The AnticipationScope will have to integrate sound, image, tactility, even smell and taste. Frequently, in PD the loss of smell precedes the characteristic tremor. Given the adaptive significance of smell in anticipatory processes, we should be able to trace changes in smell in a context of actions (such as orientation) affected by such changes.

(2) Dementia

    a. Dementia due to Alzheimer's can be distinguished from non-Alzheimer's dementia (such as fronto-temporal dementia, FTD). FTD is characterized by an early appearance of behavioral symptoms (lack of inhibition, impulsive or inappropriate behavior, outbursts) compared to AD;

b. Unlike FTD, AD patients display an early difficulty in learning and retaining new information.

Observation: Given the possibility of monitoring any system, i.e., generating a representation of its functioning (pretty much like the Anticipatory Profile that aggregates a "film" of the action and sensorial information), we could conceive of a subsystem of the AnticipationScope as a diagnostic environment for any machine. Data is collected from all components. This is the way in which the behavior of the system can be modeled. For such a system to be effective, prediction algorithms need to be developed having in mind that real-time prediction of a system's behavior is a matter of high-performance computation and of extremely efficient data mining. In 2009, years after the AnticipationScope was first made public, [15] of the California Institute of Technology came up with the notion of "*cognitive signal processing*."

6 Qualified Gazing into the Crystal Ball

The next challenge is the creation of a wearable AnticipationScope that integrates motion capture and wireless sensors in a body suit (see Figure 10).

The next step, to be carried out would entail a daunting task: to see how you feel. A colleague (Bruce Gnade, Distinguished Chair in Microelectronics, with whom I cooperate) shared the following with me:

Similar to what happened in silicon integrated circuit technology 40 years ago, flexible electronics is now at a point where system design and process integration will drive the technology. The use of an electronic textile-like architecture provides the ability to integrate several different materials and functions into the same fabric. Many of the discreet components needed for complex circuits and sensors have been demonstrated on flexible substrates: transistors, diodes, capacitors, resistors, light emitting diodes, and photo-detectors. The next step in developing the technology is to integrate these discreet components onto a single textile.

Such a wearable AnticipationScope might even become a therapeutic medium: it would not only represent the individual's anticipatory state in the form of an

Fig. 10 The wearable AnticipationScope: a possible development

Anticipatory Profile, but also allow for very sophisticated operations on such representations. Imagine: you are wearing the AnticipationScope. The 3D sensors describe your movement; the physiological and neurological sensors describe the anticipation involved. Wireless networking facilitates the real-time processing of data on high-performance computers. You can literally see an image of your own heart beating, or other functions, e.g., what happens when you are stepping, jumping, sitting down, catching a ball, hammering, designing a house, or making a model (real or virtual) of the house. Then, using biofeedback, you can reduce your heartbeat, or work on your gait and posture, or optimize your activity, etc., and immediately find out how the integrated Anticipatory Profile is affected. Accessing the integrated expression of our functioning, and trying to influence one or another variable, such as those affecting control and motoric functions, might prove to be therapeutic in more ways than we can fathom today.

Acknowledgments The presentation within the International Workshop on Next Generation Intelligent Medical Support Systems (Tîrgu Mureş, September 18–19, 2011) was made possible by the Hanse Institute for Advanced Study (Hanse Wissenschaftskolleg, Germany). The author benefited from feedback from Dr. Michael Devous (Brain Imaging, University of Texas-Southwestern Medical Center), Dr. Navzer Engineer (Neuroscience, University of Texas at Dallas), Dr. Mark Feldman (Presbyterian Hospital of Dallas). Bujor Rîpeanu provided a copy of George Marinescu's film recordings. Andres Kurismaa, a graduate student from Estonia, provided valuable insight into Bernstein's work. Irina Sukhotina facilitated contacts with the scientists in Russia researching Bernstein's legacy. Elvira Nadin provided research expertise in all the experiments, and in the various versions of this chapter. The author expresses his gratitude to Barna Iantovics an Calin Comes for their help in preparing the manuscript for print. AnticipationScope and Anticipatory Profile are trademarks belonging to Mihai Nadin.

References

1. Barboi, A., Goetz, C.G., Musetoiu, R.: MD Georges Marinesco and the early research in neuropathology. Neurology **72**, 88–91 (2009)
2. Bernstein N.: Kymocyclographic Methods of Investigating Movement, Handbuch der biologischen Arbeitsmethoden. Abt. 5. Methoden zum Studium der Funktionen der einzelnen Organe des tierischen Organismits. Teil 5a, Heft 4, Urban und Schwarzenberg, Berlin/Vienna, (1928).
3. Bernstein, N.: Essays on the physiology of movements and physiology of activity. In: Gazenko, O.G. (ed.) Series: Classics of Science. Publisher: Nauka (Science), (under the scientific direction of I. M. Feigenberg.) (1990) Book in Russian
4. Bongaardt, R., Meijer, O.G.: Bernstein's theory of movement behavior: historical development and contemporary relevance. J. Mot. Behav. **32**(1), 57–71 (2000)
5. Gallese, V.: The Inner Sense of Action: Agency and Motor Representations. J. Conscious. Stud. **7**(10), 23–40 (2000)
6. Gahery, Y.: Associated movements, postural adjustments and synergy: some comments about the history and significance of three motor concepts. Arch. Ital. Biol. **125**, 345–360 (1987)
7. Goldfarb L., Golubitsky O.: What is a structural measurement process? Faculty of computer science, University of New Brunswick, Canada. http://www.cs.unb.ca/goldfarb/smp.pdf. 30 Nov 2001

8. Latash, M.L. (ed.): Progress in Motor Control: Bernstein's Traditions in Movement Studies, Human Kinetics, vol. 1. Champaign, IL (1998)
9. Nadin, M.: MIND—Anticipation and Chaos. Belser Presse, Stuttgart/Zurich (1991)
10. Nadin M.: Anticipation and dynamics: Rosen's anticipation in the perspective of time. (Special issue) Int. J. Gen. Syst. **39**, 1 (2010) (Taylor and Blackwell, London, pp 3–33)
11. Nadin M.: Play's the thing. A wager on healthy aging. In: Bowers J.C. Bowers C (eds) Serious Game Design and Development, vol. 28(8), pp. 150–177, IGI, Hershey (2010)
12. Nadin M.: The anticipatory profile. An attempt to describe anticipation as process. Int. J. Gen. Syst. **41**(1), 43–75 (2012) (Taylor and Blackwell, London)
13. Rizzolati, G., Fadiga, L., Gallese, V., Fogassi, L.: Premotor cortex and the recognition of motor actions. Cogn. Brain Res. **3**, 131–141 (1996)
14. Rosen, R.: Anticipatory Systems. Pergamon, New York (1985)
15. Ruy de Figuieredo, J.P.: Cognitive signal processing: an emerging technology for the Prediction of behavior of complex human/machine. In: IEEE Conference on Communications, Circuits, and Systems—ICCCAS, San Jose, (2009)
16. Savage, T.: Adaptability in organisms and artifacts: A multi-level perspective on adaptive processes. Cogn. Syst. Res. Elsevier, Amsterdam **11**(3), 231–242 (2010)

Generating Sample Points in General Metric Space

László Kovács

Abstract The importance of general metric spaces in modeling of complex objects is increasing. A key aspect in testing of algorithms on general metric spaces is the generation of appropriate sample set of objects. The chapter demonstrates that the usual way, i.e. the mapping of elements of some vector space into general metric space is not an optimal solution. The presented approach maps the object set into the space of distance-matrixes and proposes a random walk sample generation method to provide a better uniform distribution of test elements.

Keywords General metric space · Sample generation · Distance cone

1 Introduction

There are several important application areas where domain objects are modeled as points in general metric space (GMS). The most widely known examples are the domains of complex pictures, words, sounds or chemical informatics [1]. The efficient management of these objects in databases is a crucial problem to be solved. There are many proposals for algorithms on storing and searching of the objects in GMS [11]. To test the cost function of the algorithms, appropriate test samples should be generated. The appropriate distribution of the test objects is a very important aspect in the evaluation of the methods.

As the objects in GMS can't be generated in a straightforward and standard way, the construction of objects in GMS is a costly and sometimes impossible task. From this reason, in most cases, the test corpus is generated as a subset of some vector space. In the Euclidean vector space, for example, the generation of vectors and coordinate values can be performed with simple and standard methods for

L. Kovács (✉)
Department of Information Technology, University of Miskolc, Miskolc, Hungary
e-mail: kovacs@iit.uni-miskolc.hu

B. Iantovics and R. Kountchev (eds.), *Advanced Intelligent Computational Technologies and Decision Support Systems*, Studies in Computational Intelligence 486, DOI: 10.1007/978-3-319-00467-9_14, © Springer International Publishing Switzerland 2014

different element distributions like uniform and normal distribution. The main goal of this chapter is to investigate the correctness of this approach. In the first part, the most important methods (Fréchet-embedding and FastMap) for mapping between GMS and the Euclidean vector space are surveyed. The performed investigations show that only a few subpart of the GMS can be covered with this projection. In order to improve the generation of the test pool, the chapter proposes a different approach for transformation between GMS and vector space. In the proposed mapping, the GMS domain elements are not the separated objects but the distance matrices of the object groups. Although, this mapping uses a higher dimensional vector space, the mapping is an injective function and it can preserve the distance relationships among the objects. The next part of the chapter investigates the measures of homogeneity and proposes a method to generate random distance matrices in this space to construct an appropriate test pool in GMS. The presented algorithm uses a Markov chain Monte Carlo walk in the high dimensional vector space. The chapter includes some test results to show the benefits of the proposed method in comparison with the traditional methods.

2 Mapping Between General Metric Space and L_2 Euclidean Space

In GMS there exists a metric on the object set to measure the distances or un-similarities between the objects. A distance function $d()$ is a metric if it meets the following conditions:

$$
\begin{aligned}
d(x,y) &\geq 0 \\
d(x,y) &= 0 \leftrightarrow x = y \\
d(x,y) &= d(y,x) \\
d(x,z) + d(z,y) &\geq d(x,y).
\end{aligned}
\tag{1}
$$

The dominating method to deal with general metric spaces is to map the objects of the GMS into a Euclidean vector space (L_2). In general case, the distance distribution in GMS cannot be preserved in L_2, i.e. no algorithm can perfectly map the objects from GMS into points of Euclidean space [2]. For example, take a distance matrix given in Table 1.

Table 1 Sample distance matrix

d_{ij}	e_1	e_2	e_3	e_4
e_1	0	5	3	2
e_2	5	0	2	3
e_3	3	2	0	2
e_4	2	3	2	0

Considering the corresponding triangles in the distance matrix representation, all of them fulfill the triangle inequality. On the other hand, it can be easily verified that it is impossible to map the objects represented by the sample matrix in Table 1 into L_2 of any dimension on a distance preserving way. It can be seen that the objects numbered with 1, 2 and 3 are situated along a line, as

$$d_{12} = d_{13} + d_{32}$$

$$\cos(\phi_{12}) = -1, \phi_{12} = \pi.$$

As a valid mapping, the objects e_1, e_2, e_3 should be mapped to points of a line segment where the distance d_{34} is equal to 1 and not 2.

As there is no distance-preserving mapping, the error of the mapping should be measured to evaluate the quality of the mapping function. The stress measure [3] shows the averaged difference between the distances in GMS and in L_2:

$$s_\mu = \sqrt{\frac{\sum_{u,v}(D(\mu(u),\mu(v)) - d(u,v))^2}{\sum_{u,v} d(u,v)^2}}$$

where

μ the mapping function
u, v objects in GMS
d distance in GMS
D distance in L_2

Another measure is the distortion that gives a threshold to the expansion of the distance values. The mapping is called non-contractive if

$$D(\mu(u),\mu(v)) \geq d(u,v).$$

The distortion [4] of a non-contractive mapping is defined as

$$a = sup_{u \neq v} \frac{D(\mu(u),\mu(v))}{d(u,v)}.$$

In this case, the following inequation holds for every element pair:

$$D(\mu(u),\mu(v)) \leq a\, d(u,v).$$

The mapping is called bi-Lipschitz [5] if there exists an $L \geq 1$ such that

$$L\, d(u,v) \geq D(\mu(a),\mu(v)) \geq \frac{1}{L} d(u,v)$$

for every (u, v) element pair. According to the results in [6], there exists embedding of GMS into L_2 with a distortion $O(logN)$, where N denotes the number of elements in GMS.

Regarding the mapping algorithm, a widely used solution is the application of the Fréchet-embedding [6]. The outline of the algorithm is the following. Let

M denote the number of the target vector space. The A_i denotes a set of pivot elements assigned to the ith coordinate. The ith coordinate of a target object u is calculated as the minimum distance value from elements belonging to A_i:

$$u_i = \mu_i(u) = min_{v \in A_i}\{d(u, v)\}.$$

Considering the L_2 metric, the distance in the vector space is calculated with the following formula:

$$D(\mu(u), \mu(v)) = \sqrt{\sum_{i=1}^{M} (d(u, a_i) - d(v, a_i))^2}.$$

In the case of MDS (multidimensional scaling) mappings [3] the initial positions are generated randomly. In the successive iterations, the positions are relocated in order to minimize the stress objective error function. The usual algorithm to solve this optimization problem is the gradient (or steepest descent) method. This method uses an analogy from physics, each pair-wise connection is treated as a spring. There is another group of variants using stochastic relocation methods [3] to find the optimal value of the objective function.

The FastMap method [7] uses an analogy of mapping from a higher dimensional space into a lower dimensional space. The projection of a point u can be calculated with the help of the Cosine Low and it yields in the following value:

$$\mu_i(u) = \frac{d(u, p_{i1})^2 - d(u, p_{i2})^2 + d(p_{i1}, p_{i2})^2}{2d(p_{i1}, p_{i2})^2}$$

where
i the index dimension,
p_{i1} the first pivot element of ith dimension,
p_{i2} the second pivot element of ith dimension

The distance values on the plane perpendicular to the selected line $p_{i1} - p_{i2}$ can be calculated with

$$d'(u, v) = \sqrt{d(u, v)^2 - (\mu_i(u) - \mu_i(v))^2}.$$

The algorithm of function FastMap consists of the following steps:

```
FP (i, D)//i - dimension index, D - distance matrix
Generate two random pivots points p_{i1},p_{i2}
Calculate μ_{i(u)} for every object
Generate the new D' distance matrix containing the d'
  distance values.
If i is less then the given dimension threshold call
  FP(i + 1, D')
else
  stop
```

It can be seen that these standard mappings from GMS into L_2 fulfils the following properties

- functions where the domain elements corresponds to objects in GMS
- not injective functions: $\exists a, b \in GMS : a \neq b \wedge f(a) = f(b)$
- the size of inverse set can be infinite : $1 \leq |f(a)^{-1}| \leq \infty$.

One of the main drawbacks of these methods is that they are not injective mappings, different object sets can me mapped to the same point set and different point set in L_2 can be mapped to the same object set in L_2.

3 Proposed Mapping of GMS into L_2

It follows from the previous considerations that it is impossible to use the inverse mapping as a function to generate a random sample in GMS. The key reason of the problem is that the mapping of points in GMS into L_2 is not an injective mapping. To cope with this problem, the proposed model uses an injective mapping where the random generated vectors can be mapped back without information loss into GMS. The key idea is that the elements of the target vector space corresponds to the distance matrices and not to the single objects.

The object set in general metric space (GMS) is described with a distance matrix $D \in \Re^{N \times N}$, where N denotes the count of elements. The matrix element d_{ij} is equal to the distance between e_i and e_j. In L_2 the point set is described with a matrix $D \in \Re^{N \times M}$, where M is the dimension of the vector space:

$$\mu : \mathcal{H} \subset \Re^{N \times N} \to \Re^{N \times M}.$$

The symbol $\mathcal{H} \subset \Re^{N \times N}$ denotes the set of distance matrices meeting the axioms of the distance functions:

$$\forall h_{ij}, h_{jk}, h_{ki} \in \mathcal{H} : h_{ij} \leq h_{ik} + h_{kj}$$
$$\forall h_{ij} \in \mathcal{H} : h_{ij} = h_{ji}$$
$$\forall h_{ii} \in \mathcal{H} : h_{ii} = 0$$
$$\forall i \neq j, h_{ij} \in \mathcal{H} : h_{ij} > 0.$$

In proposed model, the dimension of the target vector space is $\binom{N}{2}$, every pair of objects corresponds to a dimension. As the only information in GMS on objects is the distance to other objects, the distance matrix contains all information on a group of objects.

Proposition *The μ mapping is an injective but not a surjective function.*

Proof In GMS, two groups of objects are indistinguishable if the distances between the corresponding elements in both groups are the same. Let us have two different object groups in a GMS, denoted by g_1, g_2. In this case, there exists a pair of objects that have different distance values in g_1 and g_2. Thus the distance matrices are also different. Thus $g_1 \neq g_2 \Rightarrow \mu(g_1) \neq \mu(g_2)$.   □

On the other hand there can be generated such point $D \in \Re^{N \times M}$, where the values in the matrix do not meet the triangle inequality rule. Thus the function is not a surjective function.

The Fig. 1 shows that the proportion of distance matrices in $\Re^{N \times N}$ meeting the requirements of metric space is drastically decreases in dependency from the number of elements in the group (N).

Proposition *Based on the injective property, it is possible to define an inverse function from L_2 into GMS. The inverse function $\mu^{-1} : L_2 \Rightarrow \mathcal{H}$ is a bijective function and for any $\alpha \geq \frac{\max\{h_{ij}\}}{2}$, the matrix $D = \left[h_{ij}\right], \forall i \neq j : h_{ij} = \alpha$, is a valid distance matrix, i.e. $D \in \mathcal{H}$.*

Proof The μ mapping is an injective and thus the function μ^{-1} is an injective mapping as different points in L_2 denote different distance matrices. The area $\mathcal{H}$ is the set of matrices meeting the requirements of metric spaces, so it the solution set of the following system of inequations:

$$\forall i : h_{ii} = 0$$
$$\forall i,j : h_{ji} \geq 0$$
$$\forall i \neq j \neq k : h_{ik} \leq h_{ij} + h_{jk}.$$

It can be easily verified that in the case $\forall i \neq j : h_{ij} = \alpha$, all inequations are fulfilled, thus $D \in \mathcal{H}$. Another observation is that this system is a homogenous system of inequations, thus a more general statement can be set. If $x, y \in \mathcal{H}$, then

Fig. 1 Proportion of valid D in $\Re^{N \times N}$

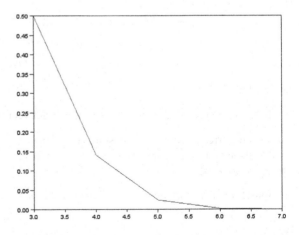

Fig. 2 Map of GMS and
euclidean points ($h_{23} = h_{14}$
$= h_{24} = h_{34} = 0.3$)

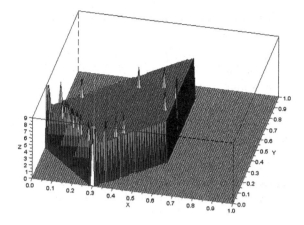

Fig. 3 Map of GMS and
euclidean points
($h_{23} = h_{24} = h_{34} = 0.4$,
$h_{14} = 0.3$)

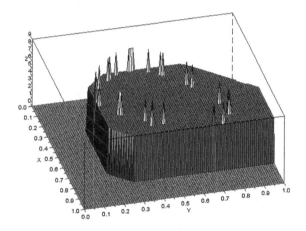

$$\forall \alpha \geq 0 : \alpha x \in \mathcal{H}$$
$$x + y \in \mathcal{H}.$$

This means that $\mathcal{H}$ is a half-subspace in $\Re^{N \times N}$.□

The Figs. 2 and 3 show some cuts of $\Re^{N \times N}$ indicating which points correspond
to valid GMS points and which ones correspond to points of L_2. The planes were
tested with brute-force testing in discrete points using a grid of 100×100. The
points with value 0, are not part of GMS, the point of value 6 or 9 are elements of
GMS. The points with value 9 are points of GMS with corresponding counterparts
in L_2.

4 Measuring the Homogeneity of Element Distribution

For generation of uniformly random set of distance matrices, a set of uniformly distributed points within the hyper-cone should be generated. The uniformity of points within a region has three main characteristics [8]:

1. the points are equally spaced,
2. the points cover the whole region,
3. the points are isotropically distributed.

The exact measuring of these properties is not a easy problem. There are several approaches to provide appropriate measures. The COV method [8] belongs to the family of point-to-point methods. The measure value α is given with

$$\alpha = \frac{1}{\bar{\gamma}} \left(\frac{1}{N} \sum_{i=1}^{N} (\gamma_i - \bar{\gamma})^2 \right)$$

where

$$\gamma_i = \min_{i \neq j} |x_i - x_j|$$

and $\bar{\gamma}$ denotes the average of minimal distance values. This method considers only the distance to the nearest point. Although it seems to be a simple method, it requires a $O(N^2)$ cost algorithm to find the nearest points. The main drawback of this measure, that it can't handle the problem of isolated (outlier) points.

The other main approach is the volumetric measure. In this case, for each points the corresponding Voronoi region is determined and the volume of the region is used to measure the uniformity. The main drawback of the method is the high computational cost to determine the membership label for every point in the region.

In our approach a distance distribution method is proposed to measure the uniformity. The distribution of the distinct distance values between the points is characteristic for the shape and dimensionality of the region. In the work [9], the expected distance distribution is calculated for a three dimensional cube. In our work, this approach is extended to high dimensional cone.

In the proposed system, the standard Euclidean ortho-normal basis is replaced with a corresponding hyperspherical coordinate system. In this coordinate system, every edge vector has a unique set of angular coordinate values. If we assume that the values of every angular coordinate may be generated from a given interval (usually from $[0, \pi]$) the cone region can be mapped to a box in the hyperspherical coordinate system.

The transformation from Euclidean coordinate $(x_1, x_2, \ldots, x_n)$ into hyperspherical coordinate $(r, \phi_1, \ldots, \phi_{n-1})$ is based on the following formulas:

$$r = \sqrt{x_n^2 + x_{n-1}^2 + \cdots + x_1^2}$$

$$\varphi_1 = arcct\left(\frac{x_1}{\sqrt{x_n^2 + x_{n-1}^2 + \cdots + x_2^2}}\right)$$

$$\varphi_2 = arcct\left(\frac{x_2}{\sqrt{x_n^2 + x_{n-1}^2 + \cdots + x_3^2}}\right)$$

$$\cdots$$

$$\varphi_{n-1} = 2 \cdot arcct\left(\frac{\sqrt{x_n^2 + x_{n-1}^2} + x_{n-1}}{\sqrt{x_n^2}}\right)$$

The other special characteristic of the proposed method is that it uses the L_1 distance instead of L_2 in order to simplify the calculations. Thus the distance between the points is measured with:

$$d(x,y) = \sum |x_i - y_i|.$$

For a one-dimensional box, the L_1 distance distribution for random points, can be calculated on the following simple way. The probability density function is equal to

$$\varphi_1(c) = \frac{2}{A}\left(1 - \frac{c}{A}\right).$$

where [0, A] denotes the interval for the coordinate values. For a two-dimensional box, the probability density function can be calculated with a convolution integral:

$$\varphi_2(c) = \int_{max(0,\,c-B)}^{min(c,\,A+B)} \varphi_1(x)\frac{2}{B}\left(1 - \frac{c-x}{B}\right)dx.$$

This formula can be extended to a higher dimension too:

$$\varphi_n(c) = \int_{max(0,\,c-A_n)}^{min(c,\,\sum_{i=1}^{n-1} A_i)} \varphi_{n-1}(x)\frac{2}{A_n}\left(1 - \frac{c-x}{A_n}\right)dx.$$

For the case when there are 5 objects and 10 dimensions, the φ_{10} was calculated using a Scilab program. The generated distance density function is shown in Fig. 4. As it can be seen the maximum density is expected at the 20 % of the maximum value.

In order to verify the theoretical calculations, also an experiment was executed. In this experiment random distance matrices were generated for $N = 5$. In the 10 dimensional space with Euclidean coordinate system, the points were generated uniformly. For the generated points, the distance distribution function was aggregated. The resulted function is shown in Fig. 5. The two distributions are similar, but having some different peak positions. The reason of the difference may be in the fact that the region of the cone is just roughly estimated.

Based on the experiments and calculations, it can be seen that the distance distribution has a very characteristic shape. The proposed measure calculates the difference between the real distance distribution on the sample set and of the theoretical distribution:

Fig. 4 The theoretical distribution

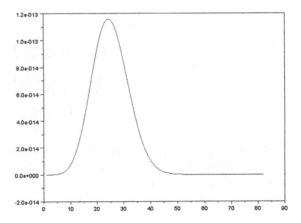

Fig. 5 The empirical distribution

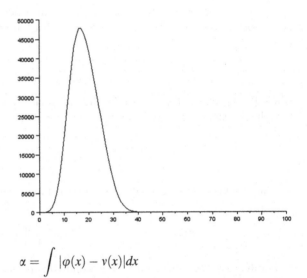

$$\alpha = \int |\varphi(x) - v(x)| dx$$

This measure requires $O(N^2)$ operations to generate the $v()$ experimental distribution, thus it is not more expensive than the traditional methods. On the other hand it considers more aspects and can handle the problem of isolated subregions too.

5 Generating Random Elements in Distance Cone

Based on the previous statements, the generation of an object set in GMS with N elements is equivalent with generating a point in $\mathcal{H} \subset \Re^{\binom{N}{2}}$. The task is equivalent with random sampling from the solution space of a system of linear inequations. In this project, the random walk approach is used to generate the random elements.

Based on [10], the methods for generating random samples in bounded regions can be clustered in four main groups: transformational methods, rejection methods, composition techniques and random walk. A transformation technique generates points in a simpler area of the vector space and maps the vectors onto the bounded target region. If the bounded region has an unregular and complex shape, the mapping function can't be generated. The composition method splits the bounded region into smaller but simpler subregions. For every subregion a special transformation mapping is used to generate the sample. In the case of rejection technique, the algorithm generates points in the whole regular space and in the second step, all generated candidate points are tested. If the candidate point lies within the bounded region, it will be inserted into the set of sample elements. In the investigated problem, this method can't be used efficiently as the probability to be inside the cone tends to zero for higher dimensions (see Fig. 3).

The proposal uses a modified version of the random walk approach [12]. The main phases of the algorithm are the followings:

- generate a starting point along the main axis of the cone,
- generate a random direction uniformly distributed in the hyperplane perpendicular to the axis,
- generate the length of the transformation vector randomly.

In the applied algorithm, the starting points are located along the main axis, i.e.

$$\forall i \neq j : h_{ij} = \text{const.}$$

The probability density of the distance from the origin is a polynomial function of the distance:

$$p(d) = \alpha d^\beta.$$

Similarly, the distance-probability from the axis is also a polynomial function. The implemented algorithm for generation of a point consists of following main steps:

Generate an α value with $\alpha > 0$ corresponding to the density $p(d)$.
Construct the initial matrix with $\forall i \neq j : h_{ij} = \alpha, h_{ii} = 0$.
Generate a direction perpendicular to main axis
Determine the maximum distance (d_m) from the axis in the selected direction
Generate a distance randomly according to density $p'_{d_m}(d)$

To incorporate the effect that the larger is α the larger is the size of a cut from the bounded region, the following distribution was selected:

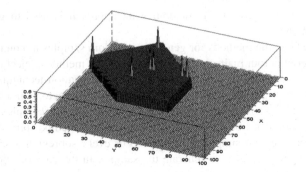

Fig. 6 Map of randomly generated points in GMS

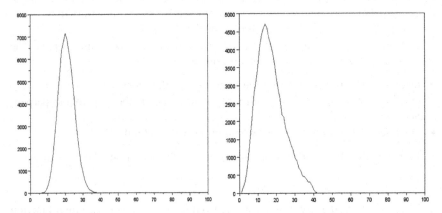

Fig. 7 Distance distribution of the euclidean mapping and of the proposed methods

$$dp(\alpha) = \frac{\alpha^k}{\int_0^{\max} \alpha^k}.$$

The generated sample set can be used as a test set with uniform distribution in GMS.

In Fig. 6, an example for random point generation is shown. In the example $N = 4$ and $h_{12} = h_{13} = h_{23} = h_{14} = 0.3$. The points with value 0 do not belong to GMS, the values with value over 0.2 are the GMS points. The points with value 0.6 are the selected elements. It can be seen that this kind of sample generation provides a better uniform distribution than the mapping from L_2 into GMS (Fig. 7). The proposed method (with $\beta = 0.8$) yields in a distribution very similar to the uniform distribution of Fig. 5.

The presented considerations and test results show that the proposed mapping between GMS and L_2 on the level of distance matrices provides an injective mapping and so it can be used as loss-less transformation between GMS and L_2. Using this invertible mapping, a more uniform sampling of objects in GMS can be achieved based on the presented random walk method. The investigation of analytical properties of the codomain of L_2 in $\Re^{N \times N}$ is an interesting question not mentioned in this chapter.

Acknowledgments This research was carried out as part of the TAMOP-4.2.1.B-10/2/KONV-2010-0001 project with support by the European Union, co-financed by the European Social Fund and the technical background was supported by the Hungarian National Scientific Research Fund Grant OTKA K77809.

References

1. Socorro, R., Mico, L., Oncina, J.: Efficient search supporting several similarity queries by reordering pivots. In: Proceedings of the IAESTED Signal Processing, Pattern Recognition and Applications, pp. 114–141 (2011)
2. Krishnamurthy, A.: Recovering euclidean distance matrix via a landmark MDS. In: Proceedings of the 6th Annual Machine Learning Symposium, October 2011
3. Kruskal, J., Wish, M.: Multidimensional Scaling. SAGE Publications, Beverly Hills (1978)
4. Abraham, I., Bartal, Y., Neiman, O.: On embedding of finite metric spaces into Hilbert space. Tech. Rep. (2006)
5. Luukkainenen, J., Movahedi-Lankarani, H.: Minimal bi-Lipschitz embedding dimension of ultrametric spaces. Fundam. Math. **144**(2), 181–193 (1994)
6. Bourgain, J.: On Lipschitz embedding of finite metric spaces in Hilbert space. Israel J. Math. **52**(1–2), 46–52 (1985)
7. Faoutsos, C., Liu, K.: FastMap, a fast algorithm for indexing, data mining and visualization of traditional and multimedia datasets. In: Proceedings of the International Conference on the Management of Data, ACM SIGMOD 1995, San Jose, CA, 163–174 (1995)
8. Gunzburger, M., Burkardt, J.: Uniformity measures for point samples in hypercubes (2004)
9. Philip, J.: The probability distribution of the distance between two random points in a box, Manuscript. Royal Institute of Technology, Stockholm (2007)
10. Smith, R.: Efficient monte carlo procedures for generating points uniformly distributed over bounded regions. Tech. Rep. 81–1, University of Michigan, USA, January, p. 21 (1981)
11. Xu, H., Agrafiotis, D.: Nearest neighbor search in general metric spaces using a tree data structure with simple heuristic. J. Chem. Inf. Comput. Sci. **43**, 1933–1941 (2003)
12. Meersche, K., Soetaert, K., Oevelen, D.: xsample(): An r function for sampling linear inverse problems. J. Stat. Softw. **30**(1), 1–15 (2009)

A Distributed Security Approach for Intelligent Mobile Multiagent Systems

Barna Iantovics and Bogdan Crainicu

Abstract Intelligent systems are used for many difficult problems solving, like: clinical decision support, health status monitoring of humans etc. The intelligence could give to a system advantages versus a system that does not have intelligence. Software mobile agents as network-computing technology has been applied for various distributed problems solving like: ubiquitous healthcare, network administration etc. The endowment of mobile agents with intelligence is difficult. Another difficulty in mobile multiagent systems consists in the limited security of mobile agents against networks sources, other agents and hosts. In this chapter a novel mobile agent architecture called *IntelligMobAg* (*Complex Intelligent Mobile Agent Architecture*) is proposed. *IntelligMobAg* represents an extension of a previous mobile agent architecture called *ICMA* [(Iantovics in A new intelligent mobile multiagent system. IEEE Computer Society Press, Silver Spring, pp. 53–159, 2005), (Iantovics in A novel mobile agent architecture. Acta Universitatis Apulensis, Alba Iulia, vol. 11, pp. 295–306, 2006)]. The purpose of the proposal consists in the limitation of some disadvantages related with security and intelligence of an *ICMA* agent by including in the static part of an evolutionary subagent. In this chapter we will present some preliminary results related with the increased security and intelligence that emerge in a cooperative multiagent system formed by agents endowed with the proposed architecture.

Keywords Computational intelligence · Intelligent system · Cooperation · Evolution · Learning · Software mobile agent · Security in mobile multiagent systems · Complex system

B. Iantovics (✉) · B. Crainicu
Department of Informatics, Petru Maior University of Tg, Mures Str. Nicolae Iorga 1, Tg. Mures, 540088 Mures, Romania
e-mail: ibarna@science.upm.ro

B. Crainicu
e-mail: cbogdan@science.upm.ro

B. Iantovics and R. Kountchev (eds.), *Advanced Intelligent Computational Technologies and Decision Support Systems*, Studies in Computational Intelligence 486, DOI: 10.1007/978-3-319-00467-9_15, © Springer International Publishing Switzerland 2014

1 Introduction

Agents and multiagent systems are appropriate for problems solving in different fields [3–7]. A cooperative multiagent system can be composed from a large number of interacting agents that as a whole have a complex problem solving behavior, which could allows the solving of some difficult problems or more efficient solving of large numbers of problems. Software mobile agents (called in the following shortly mobile agents) exhibit characteristics such as autonomy and mobility in networks [8]. Mobile agents are appropriate for many distributed problems solving, like: *ubiquitous healthcare* [9], *distributed information retrieval* [10], *support for mobile users* [10], *real-time access and control of special networked instruments* [10], *network monitoring and management* [11], *military domain* [12], *health monitoring sensor network* [13], *distributed vision sensor fusion* [14], *distributed traffic detection and management* [15], *flexible control of multi-robotic systems* [16] etc.

The protection of software mobile agents is difficult based on their distributed and asynchronous operation [17–20]. Another difficulty related with the mobile agents consists in their limited intelligence [1, 2, 21, 22]. Some times very simple efficiently and flexibly cooperating mobile multiagent systems are considered intelligent at the level of whole system [11].

In this chapter the increased intelligence and security solutions provided by a novel mobile agent architecture called *IntelligMobAg* (*Complex Intelligent Mobile Agent Architecture*) are analyzed. The proposed architecture represents an extension of a previous architecture called *ICMA* described in [1, 2].

The upcoming part of the chapter is organized as follows. In Sect. 2 security issues in mobile multiagent systems are presented; Sect. 3 presents some security solutions in mobile multiagent systems; In Sect. 4 some difficulties in the endowment of mobile agents with intelligence are presented; Sect. 5 presents the proposed *IntelligMobAg* mobile agent architecture, there are analyzed aspects related with security of a multiagent system formed by *IntelligMobAg* agents, is discussed if *IntelligMobAg* agents can form complex systems and are treated aspects related with the intelligence; In Sect. 6 the conclusions of the chapter are presented.

2 Security Issues in Mobile Multiagent Systems

The main security issues in mobile multiagent systems consist in [10, 23]: protecting the hosts against the mobile agents (mobile agents may attack the public services offered by the hosts, may launch viruses, trojan horses or worms at the hosts); protecting the mobile agents against each other (mobile agents may attack, some times modify other mobile agents that operate at the same host); protecting the mobile agents against the executing hosts (the hosts may modify the mobile agents that running on them); protecting the mobile agents during their migration against different network sources (network sources may attack or stole the mobile agents).

Some particular challenges of mobile agents consist in [8]: authentication of the agents and secrecy of the agents. The authentication of the agents consists in how can be proved that an agent it is representing whom it claims to be representing? The secrecy of the agents consists in how can be ensured that the agents maintain they owner's privacy? How do can be ensured that someone else does not read your personal agent and use it for his own gains?

It must be guaranteed that the code of a mobile agent is executed according to the specification of the middleware architecture and that of the programming language. It is important to ensure that the data and code carried by an agent is not modified during its operation or migration. The security issues in mobile multi-agent systems may be caused by errors in the executing environments or by intentional misbehavior. Such cases must be detected and prevented.

3 Security Solutions in Mobile Multiagent Systems

The security of a host consists in protecting the host from the visitor mobile agents [10]. A mobile agent at a host uses different services and resources offered by the host. However, the agent may execute different destructive actions. A malicious or erroneous mobile agent can destroy the host on which it is running if the host is inefficiently protected. Some solutions described in the scientific literature for protecting hosts against executing mobile agents consist in: *proof-carrying code* [24], *sandboxing* [25], *software fault isolation* [26] etc.

Threats against the mobile agents can be classified as those performed [10, 17, 18, 20, 27–31]: during the agents' migration and as those performed by the executing hosts. Protecting of the agents against the hosts on which they are executing is very hard [17–20]. The protection of the mobile agents is made harder by the fact that a set of malicious hosts may collaborate in the fraud. An efficient protection mechanism of a mobile agent against at a host should provide code and execution integrity (code privacy), solutions for computing with secrets (data privacy) and prevention from denial of service attacks against agents.

Mobile agents can be protected by introducing *trusted nodes* [20] into the infrastructure to which mobile agents can migrate when required (the platform from where a mobile agent is first launched has to be a trusted node). Important information can be prevented from being sent to untrusted hosts, and certain misbehaviors of malicious hosts can be traced.

Execution tracing solution proposed by Vigna [18] is appropriate for detecting unauthorized modifications of a mobile agent through recording of the agent's execution at each visited host. In this approach each host retains a log of the operations performed by the agent while executing on it. A host may execute a large number of mobile agents. A drawback of this approach consists in the size and management of all the created logs by the hosts that should be handled.

Yee [28] proposes a solution to protect the authenticity of an agent state or partial result when the agent is running at a host. Using symmetric cryptographic algorithms are generated *partial result authentication codes*. An agent is equipped with a number of encryption keys. Every time when the agent migrates from a host, the agent's state or some other result is processed using one of the keys, producing a message authentication code. The key that has been used is then disposed of before the agent migrates. The partial result authentication codes can be verified at a later point to identify certain types of tampering.

Another solution consists in encrypting the mobile agents. *Encryption* proposes that the agent's code and information are encrypted by a secret key [29]. Only a small window containing the actual point of execution would be decrypted on the fly, making possible its execution. As execution proceeds, the passed code fragments are encrypted again. A disadvantage of dynamic encryption and decryption consists in decreasing the efficiency of the agents.

Ahmed [32] proposes a protection mechanism of the mobile agents against malicious hosts called *Secure-Image Mechanism* (*SIM*). *SIM* allows protecting mobile agent by using the symmetric encryption and hash function. The mechanism proposed by Ahmed can prevent the alteration and espial attacks.

In the article [33] a protocol that protects mobile agents against malicious hosts is presented. The protocol proposed by Abdelhamid et al. combines four methods based on: reference execution (reliable platforms that shelter the proposed cooperating sedentary agents); cooperation between a mobile agent and an agent called sedentary agent. Cryptography and the digital signature ensure safe inter-agent communication and execution in deadline. A dynamic approach that makes use of a timer to make it possible to detect a mobile agent's code re-execution was used.

Karnik and Tripathi [34] propose an architecture called *Ajanta*. *Ajanta* offers some security mechanisms to protect: server resources from malicious mobile agents, agent data from tampering by malicious servers, and the system infrastructure itself. Agents' access to server resources is controlled using a proxy-based mechanism. An agent can carry three kinds of protected data: objects visible only to specific servers, read-only objects, and a secure append-only list of objects. *Ajanta* supports communication between remote agents using *java remote method invocation*. A generic authentication protocol is used for some client–server interactions.

Techniques based on *code obfuscation* (see for example a description in Gulyas et al. [10]) may secure the mobile agents. A technique based on code obfuscation proposes an extensible set of transformations to be applied to the mobile agent code. In this way, is produced code harder to read, but with identical results. This approach usually results in code with lower efficiency.

4 Intelligence in Mobile Multiagent Systems

In the following we consider traditional mobile agents described in the scientific literature. The endowment of a mobile agent with intelligence [1, 2, 21, 22], for example *evolutionary capacity*, may increase the mobile agents' body size (the code that describes the evolution must be carried by the agent in the network jointly with the code that describes problems solving) and its behavioral complexity (the code that describes the evolution must be executed at some hosts jointly with the code that describes the problem solving). The evolution usually is realized using evolutionary algorithms (some times genetic algorithms). Evolutionary algorithms may allow the learning for example. Such algorithms sometimes are called *evolutionary learning algorithms* [35]. Evolutionary algorithms are used for many problems solving [36–38]. An evolutionary learning algorithm that allows the construction of rules in the article [35] is described.

The transmission of a large number of intelligent mobile agents in a network may overload the network with data transmission. A large number of intelligent mobile agents that must be executed at a host may overload the host with data processing. The operation time of a mobile agent at a host depends on factors, like available computing resources of the hosts and overloading degree of the host. The transmission of a mobile agent can take more or less time depending on the overloading degree of the network with data transmission.

The previously mentioned motivations indicate the difficulty of establishment of the position of a mobile agent (where a mobile agent is at a moment of time). The operation of the mobile agents' members of a multiagent system is distributed and asynchronous. This makes difficult the communication and cooperation between migrating mobile agents.

In the literature [1, 2, 39–41] are presented some mobile agents considered intelligent based on different considerations. In the article [39] the mobile agents' intelligence is considered based on the autonomous and efficient migration point of view. In many researches the intelligence is considered at the level of cooperative system in that the agents operate [1, 2, 11]. Some times very simple cooperating agents can form intelligent systems (systems that intelligently solve problems). Yang et al. [11] proposes a rule-driven intelligent mobile multiagent system for real-time configuration of computer networks.

5 *IntelligMobAg* Complex Mobile Multiagent Systems

5.1 *The* IntelligMobAg *Architecture*

In the article [1, 2] a mobile agent architecture called *ICMA* have been proposed. An *ICMA* mobile agent (an agent endowed with the *ICMA* architecture) is composed from a static subagent and more mobile subagents created by the static

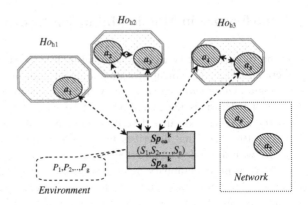

Fig. 1 Cm_k an *IntelligMobAg* agent at a problems solving cycle Cyc_z

subagent. In this chapter, for the elimination of some disadvantages of the *ICMA* agents an extended architecture called *IntelligMobAg* (*Complex Intelligent Mobile Agent Architecture*) we have proposed.

An *IntelligMobAg* agent (an agent endowed with the *IntelligMobAg* architecture) denoted in the following Cm_k is composed from two parts: a static part denoted Sp^k and a mobile part denoted Mp^k. The static part $Sp^k = \langle Sp_{oa}^k, Sp_{ea}^k \rangle$ of Cm_k is composed from an *operational subagent* denoted Sp_{oa}^k and an *evolutionary subagent* denoted Sp_{ea}^k. Sp_{oa}^k is responsible for creation of the mobile subagents at each problems solving cycle (Sp_{oa}^k creates the necessary mobile subagents according to the undertaken problems for solving). Figure 1 illustrates Cm_k at a problems solving cycle denoted Cyc_z, that consists in the solving of a set $Set = \{P_1, P_2, \ldots, P_g\}$ of undertaken problems. Mp^k at the cycle Cyc_z (see the Fig. 1) is composed from the mobile subagents $Mp_1^k, Mp_2^k, Mp_3^k, Mp_4^k, Mp_5^k, Mp_6^k$ and Mp_7^k. In Fig. 1, a_1 denotes Mp_1^k, a_2 denotes Mp_2^k, a_3 denotes Mp_3^k, a_4 denotes Mp_4^k, a_5 denotes Mp_5^k, a_6 denotes Mp_6^k and a_7 denotes Mp_7^k.

The Cyc_z problem solving cycle begins with the undertaking of a set $Set = \{P_1, P_2, \ldots, P_g\}$ of problems at the beginning of the cycle, it is followed by the creation of the set $Mp_1^k, Mp_2^k, Mp_3^k, Mp_4^k, Mp_5^k, Mp_6^k$ and Mp_7^k of mobile subagents, after then the mobile subagents are launched for problems solving. Cyc_z is finished when all the undertaken problems are solved.

The body of a created mobile subagent $Mp_i^k (Mp_i^k \in Mp^k)$ contains the agent's code (describes the solving of the undertaken problems by Mp_i^k), and may contains some data used in the problems' solving. $spec(Cm_k) = \{S_1, S_2, S_3, \ldots, S_n\}$ represents the specializations detained by Sp_{oa}^k, that allows the solving of a set *classes*$(Cl) = \{C_1, C_2, \ldots, C_n\}$ of classes of problems.

The *IntelligMobAg* architecture allows the communication of subagents of the same agent. The dashed arrows illustrated in Fig. 1 presents the communication links between the subagents of Cm_k. Sp_{oa}^k has a fixed address in the network, it does not change its location in the network during its life cycle. However, all the mobile subagents that operate at a host can communicate with Sp_{oa}^k.

The mobile subagents migrate in the network during their life cycle. Each mobile subagent communicates to Sp_{oa}^k when it arrives to a host. Each mobile subagent announces Sp_{oa}^k when it lives a host. However, Sp_{oa}^k can communicate with each created mobile subagent when this operates at a host. Two mobile subagents denoted Mp_i^k and Mp_j^k at the same host can communicate directly using a shared "blackboard memory" created at that host. Onto a backboard memory the mobile subagents must be able to read and/or write information, data and/or knowledge.

Let Mp_i^k and Mp_u^k two mobile subagents that are not at the same host (see formula 1). They should communicate using as interloper their creator operational subagent Sp_{oa}^k.

$$Mp_i^k \leftrightarrow Sp_{oa}^k \leftrightarrow Mp_u^k \tag{1}$$

In the following, we consider the scenario in that Mp_i^k intend to send a message denoted *msg* to Mp_u^k. In order to do this, Mp_i^k send *msg* to Sp_{oa}^k. Sp_{oa}^k will transmit the received message *msg* to Mp_u^k when this arrives to a host. Hovever, Mp_i^k can communicate with Mp_u^k even if Mp_u^k migrates in the network during the message transmission.

Figure 1 ilustrates a moment of time when: Mp_1^k operates at the host Ho_{h1}; Mp_2^k and Mp_3^k operate at the host Ho_{h2}; Mp_4^k and Mp_5^k operate at the host Ho_{h3}; Mp_6^k and Mp_7^k migrate in the network at that time. The subagents Mp_1^k, Mp_2^k, Mp_3^k, Mp_4^k and Mp_5^k that operate at hosts can communicate with Sp_{oa}^k. Mp_2^k and Mp_3^k can communicate with each other using the blackboard memory created by Ho_{h2}. Mp_4^k and Mp_5^k can communicate with each other using the blackboard memory created by Ho_{h3}. A migrating mobile subagent (Mp_6^k, Mp_7^k illustrated in Fig. 1) will be able to communicate with the operational static subagent when it arrives to a host. A mobile subagent at a host is able to communicate with all the mobile subagents at that host.

More *IntelligMobAg* mobile agents can form a cooperative multiagent system, that we denote in the following *CM*; $CM = \{Cm_1, Cm_2, ..., Cm_n\}$ (see Fig. 2). The motivation consists in proprieties of the *IntelligMobAg* agents like, increased: autonomy in operation, communication and cooperation capability. Let Sp^1, $Sp^2,..., Sp^n$ the static parts of the Cm_1, Cm_2, ..., Cm_n agents, where $Sp^1 = \langle Sp_{oa}^1, Sp_{ea}^1 \rangle, Sp^2 = \langle Sp_{oa}^2, Sp_{ea}^2 \rangle, ..., Sp^n = \langle Sp_{oa}^n, Sp_{ea}^n \rangle$.

In the previous paragraphs we have analyzed the communication between the subagents of the same *IntelligMobAg* agent. In the following the communication between subagents of different *IntelligMobAg* agents will be analyzed.

The operational static subagents $Sp_{oa}^1, Sp_{oa}^2, ..., Sp_{oa}^n$ can communicate directly. The communication between two mobile subagents of different *IntelligMobAg* agents, operating at the same host can be realized using a shared blackboard memory created at that host. The communication between two mobile subagents Mp_i^k and Mp_j^r (see formula 2) of different mobile agents Cm_k ($Cm_k \in CM$; Mp_i^k

Fig. 2 A cooperative
IntelligMobAg multiagent
system

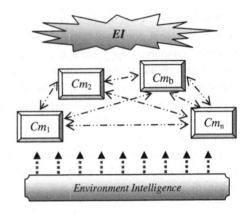

subagent of Cm_k) and Cm_r ($Cm_r \in CM$; Mp_j^r subagent of Cm_r) that operate at different hosts is realized via the operational subagents Sp_{oa}^k (Sp_{oa}^k subagent of Cm_k) and Sp_{oa}^r (Sp_{oa}^r subagent of Cm_r) of the agents (Sp_{oa}^k and Sp_{oa}^r can communicate directly).

$$Mp_i^k \leftrightarrow Sp_{oa}^k \leftrightarrow Sp_{oa}^r \leftrightarrow Mp_j^r \qquad (2)$$

Some aspects related with the intelligence in a multiagent system formed by *IntelligMobAg* intelligent agents in the Sect. 5.3 are discussed. The notation *EI* used in Fig. 2 denotes the *emergent intelligence* of the *CM* system (the intelligence considered at the level of system of agents). The agents from the system cooperatively solve problems, however, the intelligence could be considered at the systems' level. The total existent knowledge in the *CM* system is distributed between the agents $\{Cm_1, Cm_2, \ldots, Cm_n\}$ members of the system. Each operational subagent detains some problem solving knowledge (problem solving specializations) with that may endow the created mobile subagets. A mobile subagent must be endowed with all the knowledge necessary to solve the undertaken problems.

The *environment intelligence* (presented in the Fig. 2) consists in different information from that the mobile agents can learn, if they have the necessary *capability* (are able to understand and process it) and *capacity* (have the necessary computing resources). We consider that an important aspect in development of the agents is by taking into account the environment intelligence that could offer support for the agents to increase their performance in the problems solving.

5.2 Security Solutions Provided by the IntelligMobAg Architecture

In the following, we consider the cooperative *IntelligMobAg* multiagent system denoted $CM = \{Cm_1, Cm_2, \ldots, Cm_n\}$ presented in the previous section.

Security solutions at the level of an IntelligMobAg agent denoted Cmk

In formula (3) is illustrated Cm_k ($Cm_k \in CM$) at a problems solving cycle denoted Cyc_f. Cm_k is composed from the static part denoted Sp^k and the mobile part denoted Mp^k. $Mp_1^k, Mp_2^k, \ldots, Mp_n^k$ represents the created mobile subagents at Cyc_f. Sp_{oa}^k is able to use any security solution with which can be endowed a static agent (software system in generally) in the protection against different attacks. I_1, $I_2, \ldots, I_n$ represent the itineraries of $Mp_1^k, Mp_2^k, \ldots, Mp_n^k$.

$$Cm_k = \langle Sp^k \rangle + \langle Mp^k \rangle,$$
$$Sp^k = \langle Sp_{oa}^k, Sp_{ea}^k \rangle, \tag{3}$$
$$Mp^k = \{Mp_1^k[I_1], Mp_2^k[I_2], \ldots, Mp_n^k[I_n]\}$$

The description contained in the itinerary $I_h = (\{Ho_{h1}; Kb_{h1}\} \rightarrow \{Ho_{h2}; Kb_{h2}\} \rightarrow \cdots \rightarrow \{Ho_{hj}; Kb_{hj}\})$ of a mobile subagent Mp_h^k consists in the hosts Ho_{h1}, $Ho_{h2}, \ldots, Ho_{hj}$ that must be visited, specifyed in the order of their visiting. $Kb_h = \{Kb_{h1}, Kb_{h2}, \ldots, Kb_{hj}\}$ represents the problem solving knowledge used at the visted hosts by Mp_h^k. A problem solving knowledge $Kb_{hi} \in Kb_h$ may corresponds to the solving of one or more problems (at a host a mobile subagent may solve one or more subproblems).

SOL1—Self-protection mechanism by leaving the unnecessary code and data

When a mobile subagent denoted $Mp_h^k \cdot (Mp_h^k \in Mp^k)$ does not need a problem solving knowledge Kb_{hy} in the following operation at the next hosts, then Mp_h^k can leave Kb_{hy}. However, Mp_h^k launched to the next host contains less data and code that can be used, and/or modified by different malicious network sources and hosts. We denote with $|Mp_h^k(t_z)|$ the size of Mp_h^k (quantity of code and data contained in Mp_h^k body) at the moment of time t_z. We denote $|Mp^k(t_j)|$ the size of the mobile part of Cm_k (the quantity of data and code contained in all the mobile subagents) at the moment of time t_j. $|Mp_1^k(t_j)|, |Mp_2^k(t_j)|, \ldots, |Mp_n^k(t_j)|$ represents the size of mobile subagents $Mp_1^k, Mp_2^k, \ldots, Mp_n^k$ at the moment of time t_j, then is satisfied the Eq. (4).

$$|Mp^k(t_j)| = |Mp_1^k(t_j)| + |Mp_2^k(t_j)| + \cdots + |Mp_n^k(t_j)|. \tag{4}$$

Let $|Mp_h^k(t_i)|$ the size of the mobile subagent $Mp_h^k(Mp_h^k \in Mp^k)$ at the moment of time t_i ($t_i < t_{fin}$) and $|Mp_h^k(t_j)|$ the size of the mobile subagent Mp_h^k at the moment of time t_j ($t_i < t_j$ and $t_j \leq t_{fin}$); then $|Mp_h^k(t_i)| \geq |Mp_h^k(t_j)|$. $|Mp_h^k(t_i)| > |Mp_h^k(t_j)|$ if in the interval of time $[t_i, t_j]$ the subagent Mp_h^k have eliminated some of its detained problem solving knowledge. $|Mp_h^k(t_j)| = 0$ if $t_j \geq t_{fin}$ (after finishing its life cycle the size of a mobile subagent transmitted in the network became 0). $|Mp^k(t_i)| \geq |Mp^k(t_j)|$, where $|Mp^k(t_i)|$ and $|Mp^k(t_j)|$ represents the sizes of the mobile part of Cm_k at the moments of time t_i and t_j. $|Mp^k(t_i)| > |Mp^k(t_j)|$ if in the interval of time $[t_i, t_j]$ some of the mobile subagents have eliminated some of the detained problem solving knowledge. $|Mp^k(t_j)| = 0$ if all the mobile subagents have finished their life cycle at the moment of time t_j. The transmitted quantity of data and code in the mobile part Mp^k of a mobile agent Cm_k decreases in time during a problem solving cycle. However, malicious hosts and malicious network sources may receive in time less information about the mobile part of Cm_k.

SOL2—Tracking mobile subagents during their operation

When a mobile subagent $Mp_q^k\left(Mp_q^k \in Mp^k\right)$ arrives to a host Ho_d, it transmits Ho_d's address and its checksum to Sp_{oa}^k. However, Sp_{oa}^k can verify if Mp_q^k have been modified in the network during its journey from the previous host to the recent host. When a mobile subagent leaves a host Ho_d, it anounces Sp_{oa}^k about the movement, transmits the new host's address where first intend to migrate and its new checksum (the unnecessary knowledge/specializations in the following operation are deleted). The received message will allow to Sp_{oa}^k to make security verifications of the mobile subagent Mp_q^k.

Sp_{oa}^k tracks the migration of all the mobile subagents $Mp_1^k, Mp_2^k, \ldots, Mp_n^k$ created at a problem solving cycle. Hovewer, it can verify if each mobile subagent visits all the hosts specified in its itinerary. It can verify if the mobile agents are modifyed or stoled. In the case of a stolen or modified mobile subagent Mp_z^k, the operational subagent Sp_{oa}^k can take different measures. For example, Sp_{oa}^k can announces the hosts from the system to kill Mp_z^k if it arrive to them.

SOL3—Distributed operating manner of the mobile subagents

The only who knows where the mobile subagents are is Sp_{oa}^k (each mobile agent announces Sp_{oa}^k when it arrives to and lives a host). At a moment of time t_q during the problems solving cycle the mobile subagents $Mp_1^k, Mp_2^k, \ldots, Mp_n^k$ of Cm_k operate distributed at hosts or migrate in the network with the initial problem solving knowledge or a decreased knowledge, some mobile subagents have already ended their life cycle. At the same host or at the same network source there are just some of the subagents.

A host from the network can access only the subagents that it execute. A network source or a host at a moment of time t_q can access or modify only a subset Sub ($Sub \subseteq Mp^k$) of the operating mobile subagents (the mobile subagents are distributed in the network). A network source or a host doesn't have a global view about all the operating mobile subagents (it doesn't know all the data and code of the mobile subagents used in the problems solving).

Security solutions at the level of an IntelligMobAg multiagent system

In the previous paragraphs we have analysed some security solutions that implicitly emerge in case of an *IntelligMobAg* agent. In a cooperative *IntelligMobAg* multiagent system denoted $CM = \{Cm_1, Cm_2, \ldots, Cm_n\}$ in case of all of the agents $Cm_1, Cm_2, \ldots, Cm_n$ emerge the previously mentioned security solutions. Different security mechanisms described in the literature for the protection of the mobile agents can be applied for increasing of the actual security level. We have presented some security mechanisms described in the literature in Sect. 3.

The next research direction will consists in developing of a cooperative security mechanism that will allow to mobile subagents and operational static subagents to cooperate with each other in order to increase the security at the systems's level. For illustrative purposes, we consider the situation when mobile subagents cooperatively collect information about the visited hosts in order to determine if the hosts are malicious. May exists differences between information collected by different mobile subagents from the same host, hovewer the certitude about the fact that a host is malicious or not can be increased.

5.3 Complexity, Intelligence of a IntelligMobAg Multiagent System

A cooperative multiagent system composed from many *IntelligMobAg* agents that can create large numbers of mobile subagents is a *dynamic complex system*. The mobile subagents have some autonomy of migration. The control in the system is distributed, each operational static subagent partially controls the behavior of the created mobile subagents (may request the killing of sóme created mobile subagents for example). The complexity of the system from the external point of view (the party who sends the problems for solving) is not observable because is internally handled by the system. We call this complexity *internal complexity* of the system. The internal handling of the complexity by a system is very important if the system is highly complex.

The increased intelligence of an *IntelligMobAg* mobile agent and the intelligence of a cooperative *IntelligMobAg* multiagent system will be analysed in details in a future research. In the following we will illustrate just some intuitive motivations.

An *IntelligMobAg* agent denoted Cm_k can be considered intelligent based on its specific operating manner. The communication between subagents of the same agent have been previously analysed. Figure 3 ilustrates the *operational subagent* denoted Sp_{oa}^k and the *evolutionary subagent* denoted Sp_{ea}^k of the same Cm_k agent.

The autonomouse computing static subagents Sp_{oa}^k and Sp_{ea}^k operates at the same computing system (that may have any computing power with that can be endowed a computing systems), they does not migrate in the network during their operation. However, they are less limited in the utilization of computing resources than the mobile subagents. Sp_{oa}^k is able to undertake problems transmitted for solving, to create mobile subagents that must solve undertaken problems, to endow mobile subagents with the necessary problems solving specializations and launch them for problems solving to hosts distributed in the network.

Sp_{oa}^k is not able to learn. If Sp_{oa}^k receives special information that allows the learning it transmit it to Sp_{ea}^k. Sp_{oa}^k can communicate with Sp_{ea}^k using the

Fig. 3 The operational
subagent Sp_{oa}^k and the
evolutionary subagent Sp_{ea}^k of
Cm^k

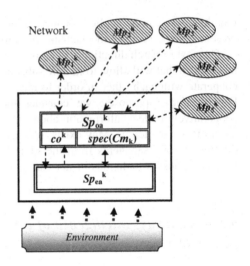

communication component denoted co^k. $spec(Cm_k)$ denotes the specializations
detained by Sp_{oa}^k. Sp_{oa}^k is able to learn, new specializations, eliminate unuseless
specializations, substitute less effective specializations with better one. This
capacity conforms to some principles called *evolutionary principles* presented in
the literature [42].

6 Conclusions

Mobile agents are very promising for many distributed problems solving. The
motivation for their limited usage consists in the limitations in security and
intelligence. Neither software system can provide a complete solution in the
protection of mobile agents [10, 20]. The limitation in intelligence [1, 2] have
practical reasons presented in Sect. 4.

In the articles [1, 2] a novel mobile agent architecture called *ICMA* have been
proposed. In this chapter an extension of this architecture called *IntelligMobAg* is
proposed. The novelty of the proposed architecture consists in a specific combination of static and mobile agent paradigms and the use of an evolutionary static
subagent. We have presented some preliminary results related with the increased
security solutions and intelligence that emerge in cooperative *IntelligMobAg*
multiagent systems. The specific operation of *IntelligMobAg* agents in a network
offer an implicit protection against other agents, hosts and network source attacks.
In a *IntelligMobAg* cooperative multiagent system the intelligence could be considered at the level of the complex dynamic system that is a problem solving
whole.

Acknowledgments The research was supported by the project "Transnational Network for Integrated Management of Postdoctoral Research in Communicating Sciences. Institutional building (postdoctoral school) and fellowships program (CommScie)"-POSDRU/89/1.5/S/63663, financed under the Sectoral Operational Programme Human Resources Development 2007–2013. The work of Bogdan Crainicu was supported by the Bilateral Cooperation Research Project between Romania and Slovakia (2011–2012) entitled: *Hybrid Medical Complex Systems*, Institute of Informatics of the Slovak Academy of Sciences and Petru Maior University.

References

1. Iantovics, B.: A new intelligent mobile multiagent system. In: Proceedings of the IEEE International Workshop on Soft Computing Applications. Szeged-Hungary and Arad-Romania, IEEE Computer Society Press, pp. 53–159 (2005)
2. Iantovics, B.: A novel mobile agent architecture, vol. 11, pp. 295–306. Acta Universitatis Apulensis, Alba Iulia (2006)
3. Iantovics, B.: Cooperative medical diagnoses elaboration by physicians and artificial agents. In: Aziz-Alaoui, M.A., Bertelle C., Cotsaftis M. (eds.) Proceedings of the Conference Emergent Properties in Natural and Artificial Complex Systems. Dresden, Germany, 2007, pp. 181–205. Le Havre University Press, Paris (2007)
4. Weiss, G.: (ed.), Multiagent systems: a modern approach to distributed artificial intelligence. MIT Press, Cambridge Massachusetts London (2000)
5. Ferber, J.: Multi-agent systems, an introduction to distributed artificial intelligence. Addison Wesley (1999)
6. Zamfirescu, C.B.: An Agent-Oriented Approach for Supporting Self-Facilitation in Group Decisions. Studies in informatics and control **12**(2), 137–148 (2004)
7. Zamfirescu, C.B., Filip, F.: Supporting self-facilitation in distributed group decisions. In: Proceedings 13th International Workshop on Database and Expert Systems Applications, DEXA'02, Aix-en-Provence, France, pp. 321–325. IEEE Computer Society Press (2002)
8. Wayner, P.: Free agents, Byte, March 105–114 (1995)
9. Kirn, St.: Ubiquitous healthcare: The onkonet mobile agents architecture. In: Aksit, M., Mezini, M., Unland, R. (eds.) Proceedings of the 3.rd International Conference Netobjectdays. Objects, Components, Architectures, Services, and Applications for a Networked World (NODe 2002), vol. LNCS 2591. Springer, Germany (2003)
10. Gulyas, L., Kovacs, L., Micsik, A., Pataki, B., Zsamboki, I.: An overview of mobile software systems, department of distributed systems. In: Computer and Automation Research Institute of the Hungarian Academy of Sciences, Mta Sztaki Technical Report TR 2000–2001 (2001)
11. Yang, K, Galis, A, Guo, X, Liu, D.Y.: Rule-driven mobile intelligent agents for real-time configuration of IP networks. , In: Palade, V and Howlett, RJ and Jain, L, (eds.) Knowledge-Based Intelligent Information and Engineering Systems, Pt 1 Proceedings, pp. 921–928, Springer, Berlin, (2003)
12. McGrath, S., Chacón, D., Whitebread, K.: Intelligent mobile agents in military command and control. Autonomous Agents Workshop (2000)
13. Chen, B., Liu, W.: Mobile agent computing paradigm for building a flexible structural health monitoring sensor network. Comput.-Aid. Civil Infrastruct. Eng., **25**(7), 504–516 (2010)
14. Nestinger, S.S., Cheng, H.H.: Flexible vision: mobile agent approach to distributed vision sensor fusion, IEEE Robot. Autom. Mag. **17**(3), 66–77 (2010)
15. Chen, B., Cheng, H.H., Palen, J.: Integrating mobile agent technology with multi-agent systems for distributed traffic detection and management systems, Transp. Res. Part C: Emer. Technol., **17**(1), February, 1–10 (2009)

16. Nestinger, S.S., Chen, B., Cheng, H.H.: A mobile agent-based framework for flexible control of multi-robotic systems, proceedings of the ASME 31th mechanisms and robotics conference, pp. 4–7. Las Vegas, Nevada (2007)
17. Sander, T., Tschudin, C.F.: Protecting Mobile agents against malicious hosts. In: Vigna G. (ed.), Mobile Agents and Security, pp. 44–60 Springer (1998)
18. Vigna, G.: Protecting mobile agents through tracing. In: Proceedings of the Third ECOOP Workshop on Operating System Support for Mobile Object Systems, Finland, pp. 137–153 (1997)
19. Nwana, H.S.: Software agents: an overview, knowledge engineering review, vol. 11(3), pp. 205–244. Cambridge University Press, Cambridge (1996)
20. Borselius, N.: Mobile agent security. Electron. Commun. Eng. J. **14**(5), 211–218 (2002)
21. Kowalczyk, R., Braun, P., Mueller, I., Rossak, W., Franczyk, B., Speck, A.: Deploying mobile and intelligent agents in interconnected e-marketplaces. J. Int. Des. Process Sci. Trans. SDPS **7**(3), 109–123 (2003)
22. Ku, H., Luderer, G.W., Subbiah, B., An intelligent mobile agent framework for distributed network management. In: Proceedings of the IEEE Global Telecommunications Conference (GLOBECOM'97), Phoenix USA, November (1997)
23. Algesheimer, J., Cachin, C., Camenisch, J., Karjoth, G.: Cryptographic security for mobile code, In: Proceedings of IEEE Symposium on Security and Privacy, pp. 2–11. Oakland, May (2001)
24. Necula, G.: Proof-carrying code. In: Proceedings of the 24th ACM SIGPLAN-SIGACT Symposium on Principles of Programming Languages, pp. 106–119. New York, Jan (1997)
25. Gong, M., Mueller, H., Prafullchandra, R. Schemers.: Going beyond the sandbox: an overview of the new security architecture in the java development kit 1.2. In: USENIX Symposium on Internet Technologies and Systems, Monterey, December (1997)
26. Wahbe, R., Lucco, S., Anderson, T., Graham, S.: Efficient software based fault isolation. In: Proceedings ACM Symposium on Operating System Principles, Dec, pp. 203–216, (1993)
27. Malik, N.S., Ko, D., Cheng, H.H.: A secure migration process for mobile agents. Softw. Pract. Experience **48**(1), 87–101 (2011)
28. Yee, B.: A sanctuary for mobile agents. In: Vitek J., Jensen C. (eds.), Secure Internet Programming, vol. LNCS 1603, pp. 261–274, Springer, New York (1999)
29. Young, A., Yung, M.: Encryption tools for mobile agents: sliding encryption. In: Biham E. (ed.): Fast Software Encryption. Proceedings of the 4th International Workshop FSE 1997, vol. LNCS 1267, pp. 230–241. Springer, Haifa (1997)
30. Pattee, H.H.: The complementary principle in biological and social structures. J Soc Biol Struct **1**, 191–200 (1978)
31. Riordan, J., Schneier, B.: Environmental key generation towards clueless agents. Vigna G. (ed.) Mobile Agents and Security, pp. 15–24 Springer (1998)
32. Ahmed, T.M.: Using secure-image mechanism to protect mobile agent against malicious hosts. World Acad. Sci. Eng. Technol. **35**, 439–444 (2009)
33. Abdelhamid, Q., Samuel, P., Hanifa, B.: A security protocol for mobile agents based upon the cooperation of sedentary agents. J. Network Comput. Appl. **30**(3), 1228–1243 (2007)
34. Karnik, N.M. Tripathi, A.R.: A security architecture for mobile agents in Ajanta, In: Proceedings of the 20th International Conference on Distributed Computing Systems, pp. 402–409 (2000)
35. Iantovics, B., Zamfirescu, C.B.: ERMS: an evolutionary reorganizing multiagent system. Int. J. Innovative Comput. Inf. Control **9**(3), 1171–1188 (2013)
36. Back, T., Fogel, D.B., Michalevitz, Z.: Handbook of evolutionary computation. Institute of Physics, Oxford University Press, New York (1997)
37. Dumitrescu, D., Lazzerini, B., Jain, L., Dumitrescu, A.: Evolutionary computing. CRC Press, Boca Raton (2000)
38. Fogel, D.B.: Evolutionary computation, toward a new philosophy of machine intelligence. IEEE Press, New York (2000)

39. Erfurth, C., Braun, P., Rossak, W.R.: Migration intelligence for mobile agents. In: proceedings of Artificial Intelligence and the Simulation of Behavior Symposium on Software Mobility and Adaptive Behavior (AISB'01), pp. 81–88. University of York, United Kingdom (2001)
40. Erfurth, C., Rossak, W.R.: Autonomous itinerary planning for mobile agents. In: Proceedings of the Third Symposium on Adaptive Agents and Multi-Agent Systems (AISB'03). pp. 120–125. University of Wales, Aberystwyth (2003)
41. Erfurth, C., Rossak, W.R.: Characterization and management of dynamical behavior in a system with mobile agents, in proceedings of the international workshop on innovative internet computing system-second, vol. LNCS, 2346, pp. 109–119. Kühlungsborn Germany (2002)
42. Pfeifer, R., Scheier, C.:. Understanding intelligence. MIT Press, September (1999)

Day Trading the Emerging Markets Using Multi-Time Frame Technical Indicators and Artificial Neural Networks

Alexandru Stan

Abstract This chapter addresses the topic of automated day trading systems based on artificial neural networks and multi-timeframe technical indicators, a very common market analysis technique. After we introduce the context of this study and give a short overview of day trading, we set out our approach and methodological framework. Then, we present the results obtained through these procedures on several of the most liquid stocks in the Romanian stock market. The final section of the chapter concludes the study and brings some insight about possible future work in the area.

Keywords Day trading systems · Neural networks · Multi-time frame technical analysis · Market forecast · Automatic trading

1 Introduction

Artificial neuronal networks have many applications [1] in the financial field as interesting and diverse as the identification of firms in financial distress, the assessment of credit rates for individuals, the detection of fraudulent behavior in bank cards usage, the forecasting of losses in insurance companies or the valuation of enterprises. In this chapter, we intend to address the topic of day trading emerging markets with artificial neuronal networks [10, 11] and multi-time frame technical indicators.

We opted for the analysis of the emerging markets [8] since many of them obviously don't satisfy the Efficient Market Hypothesis even in its weakest form, that is to say, future price movements can be predicted by using historical stock

A. Stan (✉)
Babes-Bolyai University, Cluj-Napoca, Romania
e-mail: alexandru.stan@econ.ubbcluj.ro

B. Iantovics and R. Kountchev (eds.), *Advanced Intelligent Computational Technologies and Decision Support Systems*, Studies in Computational Intelligence 486, DOI: 10.1007/978-3-319-00467-9_16, © Springer International Publishing Switzerland 2014

prices and technical analysis. If we accept the existence of these deterministic patterns, then the artificial neural networks can be put to work to find them. Naturally, the inputs for the artificial neuronal networks will be the inputs used by the technical analysts: a myriad of classical technical indicators during different time periods at different sampling frequency. Since we discuss very short term trading strategies [2] the sampling frequency will be counted in minutes rather than hours or days.

The main problem we identified when implementing trading strategies is that although some offer good accuracy over well chosen time horizons the stochastic components [2, 4] contained in financial time series may completely jeopardize their accuracy and effectiveness as the time span varies [3, 5, 9]. Thus the dynamics of an algorithmic strategy [7] may differ significantly according to the time frame we are looking at. For instance, when on a weekly basis the market may be strongly up trending while in the very short term the stock prices could be ranging between ephemeral support and resistance levels the trading automaton should adapt its behaviors to the sampling frequency. As a consequence, we decided to have a multi-timeframe approach and dedicate an autonomous artificial neural network for each time frame to identify more accurately the deterministic patterns of the corresponding sampling frequency [6]. An important characteristic when using neural networks in this context is the neural network dimensionality [12, 15], that is to say, how many steps are used to predict the best trading operations. If the dimensionality is too small the learning process may suffer and the trading strategies prove inefficient. If the dimensionality is too large an over fitting may prove even more harmful. A multiple timeframe approach can be an interesting solution since it allows for different dimensionalities in the different time frame neural networks.

2 Day Trading

This technique consists in rather making small gains by using highly leveraged transactions scaled so as to maximize financial performance. The day trader benefits from market volatility and often gains or losses on each transaction only from a 0.1 % to a few percents of the invested capital. Most often a day trader performs dozens of orders per day. All positions are closed at the end of the market session, even if losses must be taken, avoiding by this considerable overnight losses. The goal is to consistently engage in more winners than losers and ensures that losers are as small as possible.

While many trading techniques such as trend following, range trading, scalping, news trading, contrarian investing are being used by the day traders, neural network automated strategies are a good fit for intraday strategies since a well trained network can effectively detect intraday patterns, enter and exit signals and assure the short and medium term statistical consistency in non efficient emerging markets. Like any automated strategy is also eliminates any psychological bias that a human trader may present.

The main obstacle to financial effectiveness when frequently performing numerous transactions are the costs related with brokerage, commissions, spreads and slippage. For this reason, we incorporated in our strategies anti churning parameters.

3 Methodological Framework

3.1 Data Preprocessing

The inputs for the artificial neuronal networks will be the inputs used by the technical analysts: the stock prices, the traded volumes, and a plethora of technical indicators that may reveal interesting and effective in the forecasting process. These input data will have to go through a preprocessing so as to fit in the intervals [0,1] or [−1,1]. For instance, all the bounded technical oscillators having values between 0 and 100 will be scaled down by dividing their values by 100.

$$Input^t_{Oscilator} = \frac{OscillatorValue^t}{100}$$

Thus, for instance,

$$Input^t_{RSI} = \frac{RSIValue^t}{100}$$

More generally,

$$Input^t_{BoundedOscillator} = \frac{OscillatorValue^t - LowerBoundry}{UpperBoundry - LowerBoundry}$$

The unbounded oscillators are standardized if we have access to their historical values, and generally we do. After computing their estimated mean and standard deviation we can scale the input as follows:

$$Input^t_{BoundedOscillator} = \Phi\left(\frac{OscillatorValue^t - \overline{Oscillator}}{S_{oscillator}}\right)$$

where $\Phi(x) = \frac{1}{\sqrt{2\pi}}\int_{-\infty}^{x} e^{-\frac{t^2}{2}} dt$ is the normal cumulative distribution function.

If the data doesn't follow a normal distribution, statistical tests may be performed and the normal cumulative distribution function may be substituted accordingly.

For the moving average indicators, it is important to teach the neural networks not their absolute values at a given time but rather their relative dynamics (the way their first and second derivatives move) and relative values as compared to stock prices and other time moving averages. This way the ANN will be able to generalize some profitable behaviors and not be stuck with absolute and often meaningless values.

$$Input^t_{MovingAverage} = \frac{MovingAverageValue^t}{price^t * PriceScallingFactor} - \frac{1}{PriceScalingFactor}$$

The scaling factor will mainly depend on the length of time frame of the sampling and the upper boundary of the estimated variance.

$$PriceScalingFactor \approx s_{price} * \sqrt{LengthOfTheTimeframe}$$

In penny markets, characterized by huge levels of variance, the scaling factor can take big values, but in our case, since our strategies are on the very short term and the volatility moderate, scaling factors 2 turn out to be a good choice.

For the price and volume inputs we are most often interested in their dynamics than in their absolute values. That's why we use percentage changes rather than the differences of absolute values.

$$Input^t_{price} = \frac{\Delta price^t}{price^t \times PriceScalingFactor}$$

$$Input^t_{Volume} = \frac{\Delta volume^t}{volume^t \times VolumeScalingFactor}$$

If it is deemed that support and resistance levels do play an important role in improving the trading strategy then additional volume and price inputs may be added as follows:

$$Input^t_{absoluteprice} = \frac{1}{price^t}$$

$$Input^t_{absolutevolume} = \frac{1}{volume^t}$$

The percentage input prices will also help at detecting the tangent of the trend channels.

Sometimes during the preprocessing stage we also need to get rid of some basic non stochastic components such as seasonality and trends. For instance, when the trends are linear this can easily be done by first differences on the level series and when exponential by first difference on log series.

In order to get rid of complex periodic components we may use discrete time low- and high-pass frequency filters or some linear combination of them. Low- and high-pass frequency filters attenuates signals with frequencies higher and, respectively, lower than some threshold cutoff frequency. Thus, the initial sampling

Fig. 1 A 5 min low-pass
volume filtering example

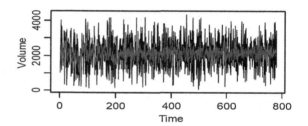

vector $\begin{bmatrix} x_1 \\ x_2 \\ x_n \end{bmatrix}$ of the prices will be transformed through filtering into a

vector $\begin{bmatrix} y_1 \\ y_2 \\ y_n \end{bmatrix}$ free of the non stochastic periodic components.

The general form of these linear filters is:

$$y_t = a_0 x_t + a_1 x_{t-1} + \cdots + a_{Dimensionality} x_{t-Dimensionality}$$

Using these filters could be quite useful in our approach based on multiple timeframes with different sampling frequencies.

This way each filter could have different dimensionalities enhancing the learning effectiveness of the corresponding artificial neural network. In a stock market time series the dimensionality stands for the number of previous pieces of information that are potentially relevant to the forecasting of the next value.

In order to determine the dimensionality of the networks in the regression models we analyzed auto correlation and partial auto correlation functions and identified past data which may cause variation in forecasting process.

The filters can also be applied on the volume data in order to focus on some aspects of the frequency domain (see Fig. 1).

3.2 The Cost of Transactions

The costs of transaction in our model will be the costs of brokerage and slippage. Since we do not always have their values at any given time we've seen them as prices, the brokerage cost as the price for a service and the slippage cost as the price for closing a position as fast as possible. Therefore we modeled them as log-normal distributions Log-N($\mu_{spread}, \sigma^2_{spread}$) and Log-N($\mu_{slippage}, \sigma^2_{slippage}$). While the spread is always there, we will suppose that the slippage cost will appear with the intensity of a homogenous Poisson process.

3.3 The Activation Function

According to the data preprocessing we used 2 activation functions. If the inputs were scaled so as to be positive we used the sigmoid function a special logistic function. Otherwise we used a hyperbolic tangent function.

If inputs $\in [0, 1]$, then the activation function is:

$$Act(x) = \frac{1}{1 + e^{-x}}$$

If inputs $\in [-1, 1]$, then the activation function is:

$$Act(x) = \frac{e^{2x} - 1}{e^{2x} + 1}$$

3.4 The Neural Networks Training

We used a bootstrapping technique to feed for instruction the different timeframe foreword neural networks. We randomly chose the beginning of the each window element in the training set. This way the training set is more homogeneous since we avoid regime changing and structural breaks in the data [13]. The downturn is that the training and validation sets can intermingle. For an element window to be valid all its points must pertain to the same daily trading session. That's why the granularity of the convolution window should rather be small.

Feeding algorithm

1. For each sampling point x_i in the data set obtained according with the sampling granularity, calculate $q_i = \frac{i}{cardinalityOfTheData}$
2. While the size of the training set is not attained do
3. Generate an random number $g \in [0, 1]$
4. If $g \in [0, q_1]$ then feed the network with the element window $[x_1, x_2, ..., x_p]$ and with the related technical indicators
5. If $g \in [q_{i-1}, q_i]$ and x_i and x_{i+p-1} belong to the same market session then feed the network with the element window $[x_i, x_{i+1}, ..., x_{i+p-1}]$ and with the related technical indicators
6. End while

3.5 Error Calculation

According to the data preprocessing and the level of financial leverage we used 3 error calculation methods. If the inputs were scaled so as to be positive and the leverage moderate we used mean squared error. If the inputs had alternating sings and the leverage was moderate we used root mean squared error. If the inputs were alternating or positives and the level of leverage was high we used arctangent root mean squared error.

We used arctangent mean squared error since it exaggerates and gives more weight to errors far away from origin and thus is useful for highly leveraged strategies.

3.6 The Networks Structure

Our data stemmed from the historical prices of the titles compounding the Bucharest Stock Exchange (BET) index for a 4 years period. We used 6 time-frames feed foreword neural networks with 5, 10, 15, 20, 25 and 30 min periods between samplings. The inputs were sliding window based. The structure of the each network for every given set of technical parameters was a multiple layer preceptor (MLP). The output layer of the MLP contains only one neuron trying to forecast the value of the stock at time $t + 1$.

After the computation of the 6 forecasting estimators obtained from the 6 neural networks (30, 25, 20, 15, 10 and respectively 5 min sampling), the strategy is to engage in a short or a long transaction only when a majority of estimates forecast the same direction for the value of the stock.

During the election process we granted more vote power to frequencies that dominated the normalized prices evolution spectrogram (Fig. 2).

Finally, as an alternative to Kelly criterion or Vince's optimal f position sizing, we resorted to an efficient frontier neural network algorithm which used all these different time frame algorithms separately as inputs for the efficient frontier combination (see Figs. 3 and 4).

3.7 Results

We used a wide range of performance and risk metrics in order to assess the performances of this technique. The most significant facts:

1. We compared the results with a set of outcomes obtained from technical indicator strategies and noticed that the algorithm had results of at least 30 % better.
2. We results were 46 % higher than a buy and hold strategy:

$$r_{\text{Buy\&Hold}} = \left(\frac{price_{sell}}{price_{buy}} \right)^{\frac{360}{daysnumber}}$$

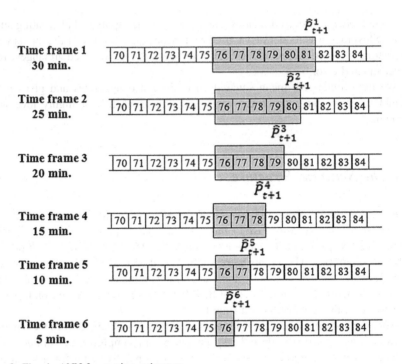

Fig. 2 The six ANN forecasting estimators

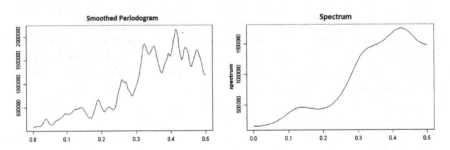

Fig. 3 Smoothed periodogram and spectrum for a share

Fig. 4 An efficient frontier combination of the 6 algorithms

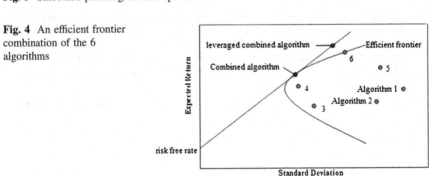

3. The maximum drawdown of the algorithm was 14 % smaller when compared with the best drawdown in the set of technical indicator strategies.
4. The length of the maximum interrupted loss was also shorter as compared to those of the other strategies.
5. We also noticed that the best performing technical indicators were those based on trend following; this is rather normal since the Hurst exponent [14] H $((R/S)t = c*t^H)$ was estimated to be around $0.624 > 0.5$, so the series were persistent and trend reinforcing.

4 Conclusions

This article confirms that technical indicators combined with neural artificial networks can produce effective trading strategies in emerging markets which exploit this inefficiency much better than the standard technical analysis strategies do. They also seem to automatically detect the type of indicators that usually best fit the market's acting mode and give them more weight and credit in the price prediction algorithms. Furthermore, the non linear structure of some ANNs provides us with adaptive filters which are more performing than standard filters applied on raw technical indicators.

Possible future evolution of the current strategy could rely upon further digital signal processing of the input technical indicators, using more elaborate filtering methods based, for instance, on Fisher and Hilbert transforms.

Acknowledgments This work is a part of CNCSIS TE_316 Grant « Intelligent Methods for Decision Fundamentation on Stock Market Transactions Based on Public Information », manager V.P. Bresfelean, Assistant Professor PhD.

References

1. Enke, D., Thawornwong, S.: The use of data mining and neural networks for forecasting stock market returns. Expert Syst. Appl. **29**(4), 927–940 (2005)
2. Farmer, D., Sidorowich, J.: Predicting chaotic time series. Phys. Rev. Lett. **59**, 845–848 (1987)
3. Fernández-Rodríguez, F., Sosvilla-Riveiro, S., García-Artile, M.: An empirical evaluation of non-linear trading rules (Working paper), FEDEA., **16** (2001)
4. Gallagher, L., Taylor, M.P.: Permanent and Temporary Components of Stock Prices: Evidence from Assessing Macroeconomic Shocks. South. Econ. J. **69**, 345–362 (2002)
5. Gately, E.: Neural networks for financial forecasting. Wiley, New York (1996)
6. de Gooijer, J.G., Hyndman, R.J.: 25 years of time series forecasting. Int. J. Forecast. **22**, 443–473 (2006)
7. Hayati, M., Shirvany, Y.: Artificial neural network approach for short term load forecasting for Illam Region. Proc. World Acad. Sci. Eng. Technol. **22** (2007)

8. Kavussanos, M.G., Dockery, E.: A multivariate test for stock market efficiency: the case of ASE. Appl. Finan. Econ. **11**(5), 573–579 (2001)

9. Lawrence, R.: Using neural networks to forecast stock market prices (1997) (Course project, University of Maritoba, Canada)

10. McNelis, D.P.: Neural networks in finance: gaining predictive edge in the market. Elsevier Academic Press, USA (2005)

11. Moreno, D., Olmeda, I.: The use of data mining and neural networks for forecasting stock market returns. Eur. J. Oper. Res. **182**(1), 436–457 (2007)

12. Priddy, L.K., Keller, E.P.: Artificial Neural Network: an introduction. SPIE Press, Washington–Bellington (2005)

13. Rotundo, G., Torozzi, B., Valente, M.: Neural networks for financial forecast, private communication (1999)

14. Qian, B., Rasheed, K.: Hurst exponent and financial market predictability, Proc. Finan. Appl. Eng. **437** (2004)

15. Walczak, S.: An empirical analysis of data requirements for financial forecasting with neural networks. J. Manage. Inf. Syst. **17**(4), 203–222 (2001)

Teaching for Long-Term Memory

Elena Nechita

Abstract A major goal of education is to help students store information in long-term memory and use that information on later occasions, in the most efficient manner. This chapter investigates the use of analogy as a strategy for encoding information in long-term memory. The results of a study concerning the ability of students to use analogy when learning computer science are presented.

1 Introduction

There are plenty of metaphors and models for memory. Starting from Aristotle, who had an integrated view on memory and thinking, human mind and its processes have been observed and studied. Nowadays, there is a huge amount of knowledge on brain, mind, and reasoning: these complex objects have been approached from different perspectives in various fields of science.

From issues addressing the evolution of the brain [21] to abstract models of the cortex [16], scientists provided insights into the mechanisms of the brain. The study of the energetics of the brain's computations is currently done, among others, by a group at the Interdisciplinary Center for Scientific Computing in Heidelberg, Germany, lead by Bert Sakmann, Nobel Prize laureate in Physiology and Medicine. With tools like algorithm-based reconstructions and annotated graphs, the scientists managed to picture the link between brain anatomy and function at the scale of tens of nanometers [7].

While biologists and neuroscientists use the modern instruments that science developed to understand the brain from the biophysics of the neuron to the biophysical basis of consciousness, from the executive functions of the human brain

E. Nechita (✉)
University "Vasile Alecsandri" of Bacău, Bacău, Romania
e-mail: enechita@ub.ro

B. Iantovics and R. Kountchev (eds.), *Advanced Intelligent Computational Technologies and Decision Support Systems*, Studies in Computational Intelligence 486, DOI: 10.1007/978-3-319-00467-9_17, © Springer International Publishing Switzerland 2014

to which sections process various types of information [5], computer scientists look at the brain in a specific manner. Namely, like a computer system who stores a great amount of information, system that we query in order to generate a desired output [17]. Using the available understanding, at a certain moment, of computers, brains and mind, we can try to understand brain by analogy, building a "cognitive computer" [18]. Even if this approach is not new, answers are still waited—can we find out more on the brain, not by using computers to study it, but modeling it as a computer system? If an end-user is not interested in *how* a computational system retrieves the information required, and looks only at *what* was returned, teachers are among those who search for answers: how can they organize "the input" such as to help students find the desired "output" as accurate as possible, in various contexts that might require their knowledge or abilities?

2 Investigation

The aim of this chapter is to point out some findings concerning the degree to which analogy can help students to have a better, long-lasting understanding of the content they are taught, and what kind of supportive elements (visual, lexical or semantic) allow them to formulate analogies. The participants of the study where 50 students (aged between 20 and 24, 86 % males, 42 % already employed) in the second year of the undergraduate programme Informatics at the Faculty of Sciences, University "Vasile Alecsandri" of Bacau, Romania. During a semester, their regular activities included the following tasks:

1. To map a real-life situation on a given totally connected graph.
2. To formulate two suggestive analogies for the database concept: one in lexical form and one in graphical form (a drawing).
3. To define (in any modality they consider appropriate) the concept of "complex system" and to give two significant examples.
4. To imagine a system analogous to the human immune system and to offer significant representations of it, in any form they find that the analogy is relevant.
5. To formulate an application of the Generalized Assignment Problem.
6. To complete phrases with analogies.
7. To integrate the same word in various lexical contexts, in order to produce different semantics.
8. To specify what elements support them for an efficient learning.
9. To describe in what way the literature they approached helped them to understand the concepts they are currently studying in computer science.
10. To present a classification type of analogy and an association type of analogy, from any topic they choose.
11. To formulate three analogies, each involving content from: literature, fine arts, and social sciences.

12. To find an analogy that includes concepts from geometry.
13. To solve a probabilities problem, given an analogous problem and its solution with Bernoulli scheme.
14. To design and implement a simulation program for the above solved problem.

The results of the students' activities have been recorded and analyzed. Moreover, discussions took place in an informal manner, aiming to reveal the needs and the perspective that the students have on the use of analogy in their future work. Students were observed during their classes (courses, seminars, laboratories) in order to register other significant aspects (such as the involvement and attention).

In what follows, findings of the study will be connected to the topics discussed.

2.1 Structure of Memory. Points at Which Information can be Lost in Learning

The understanding of such complex system as the mind clearly needs to go beyond brain structure–function correlations [26]. The way that information is organized and represented in memory has been intensively studied, but the findings remain speculative. However, according to memory researchers, the components of the Memory System are: sensory information storage (SIS), short-term memory (STM), and long-term memory (LTM). Each systems has its own characteristics with respect to function, the form of the information which can be retained, the amount of time the information is stored, and the capacity of the information that can be processed. Focusing on long-term memory, three aspects can be emphasized: episodic memory deals with the ability to recall experiences deployed in the past, stored as images; semantic memory contains verbal information, organized either as particular pieces of information (known as facts) or as generalized information (concepts, rules); procedural memory deals with the information that allows tasks performing (conditioned reflexes, emotional associations, and skills and habits).

Cognitive psychologists use an important concept related to memory organization: that of "schema". A schema denotes any pattern of relationships among data stored in memory [14]. Of course, any piece of information in memory may be connected to different, overlapping schemata. Definitional, assertional or implicational networks [25] are simple computational models of such schemata. Unlike in these semantic networks, which use a straightforward declarative representational mechanism that allows the automated reasoning, the way the output appears in the human mind is much more complex. However, like in semantic networks, the information returned as response to a query essentially depends on how previous knowledge is connected in memory and on how the pieces are used to build this output. Accordingly, the content that already exists in students' minds and the understanding that they got on that content have a major influence on what students integrate or experience as new content. They can retrieve only their interpretation on what they stored in long-term memory, as far as mental models cannot be avoided.

An important aspect related to the schemata is the fact that these are resistant to change. Meanwhile, new mind-sets tend to form quickly—that is why teachers need to facilitate for their students a meaningful context that includes at least: examples, analogies, and alternative interpretations. In the rapidly changing environment that we all face today, this should not be a difficult task. Previous, well-organized pieces of knowledge, past experiences, cultural values, as well as the stimuli perceived by the students in the learning space have a major influence on their new acquisitions.

For the study group, it appears that the transfer of the information in the short-term memory (first point where problems may appear) suffers from lack of attention and, when attention is given to the current activity, from a superficial level of attending it. From the questionnaires and from our direct observations, this attitude comes from a weak motivation or even from its absence. There are multiple sources of such an attitude, the top two being: a bad understanding of what a computer scientists should learn (for example, students share the opinion that the mathematical instruments are not important, or that the weight of the theoretical lessons is too big), and a perspective that they do not appreciate as favorable (due to the actual economical context; some of the graduates happen to work in other fields of activity). However, those who are interested in the topics approached (and who have no deficits in sensory systems such as visual, auditory or kinesthetic) mentioned a friendly, informal, quiet environment, music, and breaks as factors that allow them to elaborate on the incoming information. Some of them also mentioned personal good will as crucial in their implication.

In this stage, the teacher needs to pay attention to the following aspects: to present the information clearly (so that it reaches the sensory register and is correctly perceived by the students) and ensure that students attend the information, focusing on relevant aspects and not on collateral ones.

A second point where problems can arise is working memory, where the information must be held long enough to work with it. When approaching a problem in computer science, several components must be referred in order to solve it: data structures, algorithms, heuristics, strategies, programming languages, complexity issues, etc. First of all, the teacher must help the students to identify these pieces of knowledge. Activating them, as has been proven in physiology [11], facilitates the transfer to long-term memory, as well as bringing it back for future use. Another role of the teacher is to provide support for prior information that is needed. Considering this aspect, the study group pointed out that students appreciate that written information (on paper) is preferred when working on a problem, as well as the process of rewriting it. Also, this is the point where mastering the basic skills appears to be critical, because their use must not occupy space in the working memory.

The third point at which students may not properly deal with information is the bridge from working memory to long-term storage. The success of this transfer is related to the following important aspects: development of associations between the new input and the schemata already existing in the memory, the work with the information in more than a single context, and the complexity level of these processes. Categorization (which is the capacity to place new information in several categories, in order to create multiple pathways to access it) appeared to be one of the problems faced by the students in the study group. In the applied questionnaires, the 10 items dealing with categorization (each item appreciated with 0 up to 5 points, according to the correctness of the answer) scored a minimum of 3 and a maximum of 38 points, the average score being only 21.4. Meanwhile, the 10 items measuring the amount of effort and cognitive capacity used to process information (the complexity level of processing; each item also appreciated from 0 to 5, according to the complexity and the completeness of the answer) scored a minimum of 2 and a maximum of 40 points, the average score being only 18.75.

As a conclusion with important consequences in education, it appears that memory storage is an ongoing process resulting from continuous changes and parallel processing in the brain [10]. Therefore, increasing the number of examples, the variability of examples, and the use of those that minimize the cognitive load [27] can improve schemata acquisition.

In the matter of memory retrieval (the fourth critical point where problems can appear in memory processing), medical studies [19] proved that, in accordance with the tasks to be solved, a selection between competing representations is carried out, in interaction with domain-general cognitive control. The following section considers analogy as a system of representation and its role in memory retrieval, problem solving and understanding, in general.

2.2 Use of Analogy in Learning and Retrieval of Information

At humans, learning is a process indistinguishable from evolution itself [3]. Learning through analogy is a superior form of learning. In fact, analogy has a major role in our lives: starting with basic things that we learn by imitation in childhood, going through the use of language to the completion of skills and competencies in our professional development, most of the learning acts use various forms of analogy.

According to Cambridge Advanced Learner's Dictionary and Thesaurus, analogy denotes "a comparison between things which have similar features, often used to explain a principle or idea". Collins English Dictionary (11th edition) explains analogy as "agreement or similarity, especially in a certain limited number of features or details" for two relational systems. Human mind, as well as human intelligence and language proved to be fundamentally analogical and figurative [28]. The history of science counts numerous discoveries that are due to analogies. Creative scientists' statements prove that new theories appeared when they applied, in their field of expertise, an analogy to a phenomenon observed in another domain, sometimes very disparate [6, 29].

The use of analogy in science has been widely investigated. Philosophy [1], rhetoric [20], linguistics [2], cognitive psychology [12], pedagogy [4] expressed their perspectives on analogy and its uses. During the last three decades, new insights into the role of analogy have also been provided by numerous computational models (see [8] for a comprehensive list, also [9]).

Because most of the projects in computer science relate to complex problem-solving tasks, our investigation also focused on: the ability of the students to recognize similarities between problems belonging to separate domains, their capacity to transfer solutions from one problem to another, how they represent analogy, and the predominant modality they use to represent analogical relationships (through text or visual image).

The process of problem solving has been magnificently analyzed by Polya in "How to Solve It. A new aspect of mathematical method" [22], for the world of mathematics. But the main idea of this book, which can be generalized to all fields, idea that is important for students and teachers as well, is that learning must be active.

Basically, the process of analogical problem solving involves three steps [13]: a representation of the original and of the target problem (1), mapping of the two representations (2), and use of the mapping to generate the solution to the target problem (3). What can a teacher do in order to constantly improve the ability of its students to operate successfully on all three steps?

The first stage-representation—is crucial. Basically, it decides how difficult the solution is to be found. Students in Informatics are presented several strategies to build representations in the Artificial Intelligence course, where analogy is intensively used. The students in the sample group did not attend this course at the moment of the study; therefore, their knowledge on representations comes from the background. The items designed to reflect their capacity to build representations invoked previous knowledge, acquired in fundamental courses and programming practice. But, most of all, those items intended to make students use their creativity, to rethink objects and situations, to find new ideas that might work on the indicated topics, hence avoiding "functional fixedness" [24]. The results, however, denoted that imagination is not one of the students' strong points: only standard, common representations have been indicated. The average score of the items dealing with representation ability was 20.35 (on a scale from 0 to 50). Descriptive representations (although imperfect and suffering from lack of details) predominated over the visual ones (very few students chose to give graphic representations: 6 for task 2 and 8 for task 4; the solutions provided by students used mainly graphs and data flow graphs). The interface designed for the simulation program required for task 14 was a graphical one in only 7 cases (representing 38.8 % from a total of 18 functional applications and only 14 % of all 50 expected cases). On task 12, only 16 correct answers (32 %) were recorded. Interviews revealed that experience and intuitive, specific examples are considered by students as elements that can improve their capacity to frame good representations.

The second stage in analogical problem solving—mapping of the representations of two similar systems—is fundamental in applications and in several superior courses (such as Multiagent Systems and Natural Computing, for example). Analogical mapping is the process of determining the best correspondence between the objects and relations of an original and of a novel, target problem. If the students are given the two systems, their corresponding elements are easily recognized, in general. On the same scale (0–50), the average score of the items dealing with mapping of representations was 28.7. But path-mapping theory [23] pointed out on the integration of analogical mapping in the process of problem solving. As the following paragraph shows, this integration has not been achieved by the students in the study group.

The third step (use of the mapping to find a solution for the target problem) was not as successful as the previous one. The average score for this step was 22.4 (of a possible maximum of 50, corresponding to task 13). The explanations that we have found for this result is that, in general, our students tend to develop a lack of persistency in working for a goal. Also, the concepts and results required by this particular analogy were not sufficiently familiar to the students.

The items dealing with lexical contexts and semantics attained to an average score of 20.5 (out of 50). Corresponding to task 11, relevant analogies have been formulated as follows: 25 in social sciences, 32 in literature, and only 14 in fine arts (representing, accordingly: 50, 64, and 28 % of the expected answers). Students described that some pieces of literature they have read helped them to understand a few concepts in computer science, also that examples coming from real life situations and projections in specific application areas remain for a long time in their memory. For example, an application in medicine that they have been presented [15] has been mentioned by students related to task 3, in 5 cases.

3 Conclusions

The study allowed us to extract some practices which might help students to perform better when they need to access information acquired long ago. The following are to be experienced by the students: over-learn new material—in order to imprint the information; read actively—to improve short-term memory registration; explain and communicate on the problem to be solved—this could activate some mind maps; do research on the problem (reading, thinking)—some analogous problems may appear, previous experiences may come in mind; approach a holistic perspective—do not limit the problem, search for as many applications as possible; make connections to other domains—even if only weak, on the surface links with the problem are seen; keep in mind the goal—the solution of the problem; summarize information from various sources—this could offer new perspectives or details; when solving the problem, have initiatives—take decisions and evaluate them; put everything on paper—this will help working memory; be creative, have the courage to do things—there is an inherent uncertainty in problem solving; have a positive attitude and perseverance.

Teachers can help their students to enhance long-term memory and to develop a more effective memory with various strategies: encourage and demonstrate the construction of mental images that store the important aspects of the problem; avoid teaching algorithmically; compare a few analogies during teaching; facilitate retrieval practice (reviewing information, tests) and learning from feedback; give the students opportunity and space to reflect on what is presented; create them various scenarios to recall and apply the newly-formed memories; ensure that students store accurate information; when designing lessons, pay attention to the time needed for consolidation; assign a meaning for the educational process.

All these attitudes might improve students' subsequent retrieval of information. Some have been particularly experienced on problem solving in computer science, but most of them are consistent with every instruction field. The use of "brain-friendly" techniques, examples, simulations, role playing [30] can significantly improve students' learning.

In terms of an overall assessment of the conducted experiment, we consider that the students in the study group have to improve their skills in working with analogy, as well as in other general aspects: reading, building vocabulary, time management, up to developing a personal learning style.

Acknowledgments This research was supported by the project entitled *Hybrid Medical Complex Systems—ComplexMediSys* (2011–2012), a bilateral research project between Romania and Slovakia.

References

1. Ashworth, E.J.: Signification and modes of signifying in thirteenth-century logic: a preface to aquinas on analogy. Medieval Philosophy Theology **1**, 39–67 (1991)
2. Anderson, J.M.: Structural analogy and universal grammar. Lingua **116**(5), 601–633 (2006)
3. Atmar, W.: Notes on the simulation of evolution. IEEE Trans. Neural Networks **5**(1), 130–147 (1994)
4. Aubusson, P.J., Harrison, A.G., Ritchie, S.M. (eds.): Metaphor and analogy in science education. Springer (2006)
5. Bear, M.F., Connors, B.W., Paradiso, M.: Neuroscience: exploring the brain, 3rd ed. Baltimore, MD: Lippincott, Williams and Wilkins (2006)
6. Boden, M.: The creative mind: myths and mechanisms, 2nd edn. Routledge, London (2004)
7. Boudewijns, Z.S., Kleele, T., Mansvelder, H.D., Sakmann, B., de Kock, C.P., Oberlaender, M.: Semi-automated three-dimensional reconstructions of individual neurons reveal cell type-specific circuits in cortex. Commun Integrative Biol **4**(4), 486–488 (2011) doi:10.4161/cib.4.4.15670
8. Davies, J., Nersessian, N.J., Ashok, K.G.: Visual models in analogical problem solving. Found. Sci. **10**, 133–152 (2005)
9. Farrell, S., Lewandowsky, S.: Computational modeling in cognition: principles and practice. SAGE Publications (2011)

10. AnA, Fingelkurts, AIA, Fingelkurts: Persistent operational synchrony within brain default-mode network and self-processing operations in healthy subjects. Brain Cogn. **75**(2), 79–90 (2011)
11. Fusi, S.: Long term memory: encoding and storing strategies of the brain. Neurocomputing **38–40**, 1223–1228 (2001)
12. Gentner, D., Holyoak, K.J., Kokinov, B. (eds.): The analogical mind: perspectives from cognitive science. MIT Press, Cambridge (2001)
13. Gick, M.L., Holyoak, K.J.: Analogical problem solving. Cogn. Psychol. **12**, 306–355 (1980)
14. Heuer, R.J.: Psychology of intelligence analysis. Center for the Study of Intelligence, USA (1999)
15. Iantovics, B.: Agent-based medical diagnosis systems. Comput. Inf. **27**(4), 593–625 (2008)
16. Johansson, C.: An attractor memory model of neocortex, Ph. D. Thesis, (2006) ISBN 91-7178-461-6, ISSN-1653-5723, ISRN-KTH/CSC/A–06/14–SE, School of Computer Science and Communication, Royal Institute of Technology, Sweden
17. Johnson-Laird, P.N.: The computer and the mind. Harvard University Press, Cambridge Mass (1988)
18. Kanerva, P.: Dual role of analogy in the design of a cognitive computer. In: Holyoak, K.J., Gentner, D., Kokinov, B.N. (eds.) Advances in analogy research: integration of theory and data from the cognitive, computational, and neural sciences, pp. 164–170. New Bulgarian University Press, Sofia (1998)
19. Kompus, K.: How the past becomes present: neural mechanisms governing retrieval from episodic memory. University dissertation from Umea (2010) http://www.dissertations.se/dissertations/311a22c472/
20. Little, J.: Analogy in science: where do we go from here? Rhetoric Soc. Q. **30**(1), 69–92 (2000)
21. Northcutt, R.G.: Understanding vertebrate brain evolution. Integr. Comp. Biol. **42**(4), 743–756 (2002)
22. Polya, G.: How to solve it. A new aspect of mathematical method. Princeton University Press, New Jersey (1957)
23. Salvucci, D.D., Anderson, J.R.: Integrating analogical mapping and general problem solving: the path-mapping theory. Cognitive Sci. **25**, 67–110 (2001)
24. Solomon, I.: Analogical transfer and functional fixedness in the science classroom. J. Educ. Res. **87**, 371–377 (1994)
25. Sowa, J.F. (ed.): Principles of semantic networks: explorations in the representation of knowledge. Morgan Kaufmann Publishers, San Mateo (1991)
26. Stephan, K.E., Riera, J.J., Deco, G., Horwitz, B.: The brain connectivity workshops: moving the frontiers of computational systems neuroscience. NeuroImage **42**, 1–9 (2008)
27. Sweller, J.: Cognitive load during problem-solving: effects on learning. Cognitive Sci. **12**(2), 257–285 (1988)
28. Turner, M.: The literary mind. Oxford University Press, New York (1996)
29. Vosniadou, S., Ortony, A. (eds.): Similarity and analogical reasoning. Cambridge University Press (1989)
30. Wolfe, P.: Brain matters: translating research into classroom practice, 2nd edn. Association for Supervision and Curriculum Development, Alexandria (2010)

A Recognition Algorithm and Some Optimization Problems on Weakly Quasi-Threshold Graphs

Mihai Talmaciu

Abstract Graph theory provides algorithms and tools to handle models for important applications in medicine, such as drug design, diagnosis, validation of graph-theoretical methods for pattern identification in public health datasets. In this chapter we characterize weakly quasi-threshold graphs using the weakly decomposition, determine: density and stability number for weakly quasi-threshold graphs.

1 Introduction

The well-known class of cographs is recursively defined by using the graph operations of 'union' and 'join' [1]. Bapat et al. [2], introduced a proper subclass of cographs, namely the class of weakly quasi-threshold graphs, by restricting the join operation. The class of cographs coincides with the class of graphs having no induced P_4 [3]. Trivially-perfect graphs, also known as quasi-threshold graphs, are characterized as the subclass of cographs having no induced C_4, that is, such graphs are $\{P_4, C_4\}$-free graphs, and are recognized in linear time [4, 5]. Another subclass of cographs are the $\{P_4, C_4, 2K_2\}$-free graphs known as threshold graphs, for which there are several linear-time recognition algorithms [4, 5]. Every threshold graph is trivially-perfect but the converse is not true.

When searching for recognition algorithms, frequently appears a type of partition for the set of vertices in three classes A, B, C, which we call a *weakly decomposition*, such that: A induces a connected subgraph, C is totally adjacent to B, while C and A are totally nonadjacent.

The structure of the chapter is the following. In Sect. 2 we present the notations to be used, in Sect. 3 we give the notion of weakly decomposition and in Sect. 4

M. Talmaciu (✉)
Department of Mathematics and Informatics, "Vasile Alecsandri" University of Bacău, Str. Calea Marasesti No 157, 600115 Bacău, Romania
e-mail: mtalmaciu@ub.ro

B. Iantovics and R. Kountchev (eds.), *Advanced Intelligent Computational Technologies and Decision Support Systems*, Studies in Computational Intelligence 486, DOI: 10.1007/978-3-319-00467-9_18, © Springer International Publishing Switzerland 2014

we give a recognition algorithm and determine the clique number, the stability number on weakly quasi-threshold graphs.

2 General Notations

Throughout this chapter, $G = (V, E)$ is a connected, finite and undirected graph, without loops and multiple edges [6], having $V = V(G)$ as the vertex set and $E = E(G)$ as the set of edges. $\overline{G}$ is the complement of G. If $U \subseteq V$, by $G(U)$ we denote the subgraph of G induced by U. By $G - X$ we mean the subgraph $G(V - X)$, whenever $X \subseteq V$, but we simply write $G - v$, when $X = \{v\}$. If $e = xy$ is an edge of a graph G, then x and y are adjacent, while x and e are incident, as are y and e. If $xy \in E$, we also use $x \sim y$, and $x \nsim y$ whenever x, y are not adjacent in G. A vertex $z \in V$ distinguishes the non-adjacent vertices $x, y \in V$ if $zx \in E$ and $zy \notin E$. If $A, B \subset V$ are disjoint and $ab \in E$ for every $a \in A$ and $b \in B$, we say that A, B are *totally adjacent* and we denote by $A \sim B$, while by $A \nsim B$ we mean that no edge of G joins some vertex of A to a vertex from B and, in this case, we say that A and B are *non-adjacent*.

The *neighbourhood* of the vertex $v \in V$ is the set $N_G(v) = \{u \in V : uv \in E\}$, while $N_G[v] = N_G(v) \cup \{v\}$; we simply write $N(v)$ and $N[v]$, when G appears clearly from the context. The neighbourhood of the vertex v in the complement of G will be denoted by $\overline{N}(v)$.

The neighbourhood of $S \subset V$ is the set $N(S) = \cup_{v \in S} N(v) - S$ and $N[S] = S \cup N(S)$. A *clique* is a subset Q of V with the property that $G(Q)$ is complete. The *clique number* or *density* of G, denoted by $\omega(G)$, is the size of the maximum clique. A clique cover is a partition of the vertices set such that each part is a clique. $\theta(G)$ is the size of a smallest possible clique cover of G; it is called the *clique cover number* of G. A stable set is a subset X of vertices where every two vertices are not adjacent. $\alpha(G)$ is the number of vertices is a stable set o maximum cardinality; it is called the *stability number* of G. $\chi(G) = \omega(\overline{G})$ and it is called *chromatic number*.

By P_n, C_n, K_n we mean a chordless path on $n \geq 3$ vertices, a chordless cycle on $n \geq 3$ vertices, and a complete graph on $n \geq 1$ vertices, respectively.

A graph is called *cograph* if it does not contain P_4 as an induced subgraph.

Let $\mathscr{F}$ denote a family of graphs. A graph G is called $\mathscr{F}$-*free* if none of its subgraphs is in F.

3 Preliminary Results

3.1 Weakly Decomposition

At first, we recall the notions of weakly component and weakly decomposition.

Definition 1 [7–9] A set $A \subset V(G)$ is called a weakly set of the graph G if $N_G(A) \neq V(G) - A$ and $G(A)$ is connected. If A is a weakly set, maximal with respect to set inclusion, then $G(A)$ is called a weakly component. For simplicity, the weakly component $G(A)$ will be denoted with A.

Definition 2 [7–9] Let $G = (V, E)$ be a connected and non-complete graph. If A is a weakly set, then the partition $\{A, N(A), V - A \cup N(A)\}$ is called a weakly decomposition of G with respect to A.

Below we remind a characterization of the weakly decomposition of a graph. The name of "*weakly component*" is justified by the following result.

Theorem 1 [8–10] *Every connected and non-complete graph* $G = (V, E)$ *admits a weakly component* A *such that* $G(V - A) = G(N(A)) + G(\overline{N}(A))$.

Theorem 2 [8, 9] *Let* $G = (V, E)$ *be a connected and non-complete graph and* $A \subset V$. *Then* A *is a weakly component of* G *if and only if* $G(A)$ *is connected and* $N(A) \sim \overline{N}(A)$.

The next result, that follows from Theorem 1, ensures the existence of a weakly decomposition in a connected and non-complete graph.

Corollary 1 *If* $G = (V, E)$ *is a connected and non-complete graph, then* V *admits a weakly decomposition* (A, B, C), *such that* $G(A)$ *is a weakly component and* $G(V - A) = G(B) + G(C)$.

Theorem 2 provides an $O(n + m)$ algorithm for building a weakly decomposition for a non-complete and connected graph.

Algorithm for the weakly decomposition of a graph ([10])
Input: A connected graph with at least two nonadjacent vertices, $G = (V, E)$.
Output: A partition $V = (A, N, R)$ such that $G(A)$ is connected, $N = N(A)$, $A \not\sim R = \overline{N}(A)$.
begin
        $A :=$ any set of vertices such that
        $A \cup N(A) \neq V$
        $N := N(A)$
        $R := V - A \cup N(A)$
        while $(\exists n \in N, \exists r \in R$ such that $nr \notin E$) *do*
            begin
                $A := A \cup \{n\}$
                $N := (N - \{n\}) \cup (N(n) \cap R)$
                $R := R - (N(n) \cap R)$
            end
end

In [7] we give:
Let $G = (V, E)$ *be a connected graph with at least two nonadjacent vertices and* (A, N, R) *a weakly decomposition, with A the weakly component. G is a* P_4-free *graph if and only if:*

(1) $A \sim N \sim R$;

(2) $G(A)$, $G(N)$, $G(R)$ are P_4-free graph.

3.2 Weakly Quasi-Threshold Graphs

In this subsection we remind some results on weakly quasi-threshold graphs.

A *cograph* which is C_4-free is called a *quasi-threshold* graph.

In [2] we study the class of weakly quasi-threshold graphs that are obtained from a vertex by recursively applying the operations (1) adding a new isolated vertex, (2) adding a new vertex and making it adjacent to all old vertices, (3) disjoint union of two old graphs, and (4) adding a new vertex an making it adjacent to all neighbours of an old vertex.

Let $G = (V, E)$ be a graph. Define a relation on V [2] as follows: Let $u, v \in V$. Then $u \equiv v$ if $N(u) = N(v)$. We observe that $\equiv$ is an equivalence relation and the equivalence classes are stable sets in G.

Let G be a graph with $Q_1, ..., Q_k$ as the equivalence classes under the relation $\equiv$. For each $i = 1, ..., k$ choose a vertex $u_i \in Q_i$. We call the subgraph $\widetilde{G}$ of G induced by $u_1, ..., u_k$ as a subgraph of representatives of G.

Let G be a graph. Then G is weakly quasi-threshold [2] if an only if a subgraph of representatives is quasi-threshold.

Let $G = (V, E)$ be a connected graph. Then the following are equivalent [2]:

(1) G is a weakly quasi-threshold

(2) G ia a P_4-free and there is no induced $C_4 = [v_1, v_2, v_3, v_4]$ with $N(v_1) \neq N(v_3)$ and $N(v_2) \neq N(v_4)$.

A graph G is weakly quasi-threshold [11] if and only if G does not contain any P_4 or $co - (2P_3)$ as induced subgraphs.

4 New Results on Threshold Graphs

4.1 Characterization of a Weakly Quasi-Threshold Graph Using the Weakly Decomposition

In this paragraph we give a new characterization of weakly quasi-threshold graphs using the weakly decomposition.

Theorem 3 *Let $G = (V, E)$ be a connected graph with at least two nonadjacent vertices and (A, N, R) a weakly decomposition, with A the weakly component. G is a weakly quasi-threshold graph if and only if:*

(1) $A \sim N \sim R$;

(2) $G(N)$ is $\overline{P}_3$-free graph;

(3) $G(A \cup N)$, $G(N \cup R)$ are weakly quasi-threshold graphs.

Proof Let $G = (V, E)$ be a connected, uncomplete graph and (A, N, R) a weakly decomposition of G, with $G(A)$ as the weakly component.

At first, we assume that G is weakly quasi-threshold. Then G is P_4-free. So, $A \sim N \sim R$. Because G is weakly quasi-threshold graph it follows that $G(A \cup N)$, $G(N \cup R)$ are weakly quasi-threshold graphs. We suppose that $G(N)$ contain $\overline{P}_3 = (\{a, b, c\}, \{ac\})$ as induced subgraph. Because $G(A)$ is connected $\exists x, y \in A$ such that $xy \in E$. Because $A \nsim R$, $\forall z \in R$, $G(\{x, y, z\}) \simeq \overline{P}_3$. Because $N \sim A \cup R$, $G(\{a, y, a, b, c, z\}) \simeq co - (2P_3)$, in contradicting with G is weakly quasi-threshold graph.

Conversely, we suppose that (1), (2) and (3) hold. From (3), $G(A)$, $G(N)$, $G(R)$ are P_4-free. Because (1) hold, G is P_4-free. $G(A)$, $G(N)$, $G(R)$ are $\{co - (2P_3)\}$-free because (3) hold. $G(A \cup R)$ is $\{co - (2P_3)\}$-free because $A \nsim R$ and $\{co - (2P_3)\}$ is connected. We suppose that G contain $H = \{co - (2P_3)\}$ as induced subgraph such that $V(H) \cap A \neq \emptyset$, $V(H) \cap N \neq \emptyset$ and $V(H) \cap R \neq \emptyset$. Because (1) hold, $N \sim (A \cup R)$. The unique $S \subset V$ totally adjacent with $V(H) - S$, $(S \sim V(H) - S)$, is S with $S = V(\overline{P}_3)$. Then $G(N)$ contain $\overline{P}_3$ as induced subgraph, contradicting (2). So, G is $\{co - (2P_3)\}$-free. So, G is weakly quasi- threshold graph.

4.2 Determination of Clique Number and Stability Number for a Weakly Quasi-Threshold Graph

In this paragraph we determine the stability number and the clique number for weakly quasi-threshold graphs.

Proposition 1 *If $G = (V, E)$ is a connected graph with at least two nonadjacent vertices and (A, N, R) a weakly decomposition with A the weakly component then*

$$\alpha(G) = max\{\alpha(G(A)) + \alpha(G(\overline{N}(A))), \alpha(G(A \cup N(A)))\}.$$

Proof Indeed, every stable set of maximum cardinality either intersects $\overline{N}(A)$ and in this case the cardinal is $\alpha(G(A)) + \alpha(G(\overline{N}(A)))$ or it does not intersect $\overline{N}(A)$ and has the cardinal $\alpha(G(A \cup N(A)))$.

Theorem 4 *Let $G = (V, E)$ be connected with at least two non-adjacent vertices and (A, N, R) a weakly decomposition with A the weakly component. If G is a weakly quasi-threshold graph then*

$$\alpha(G) = \alpha(G(A)) + max\{\alpha(G(N)), \alpha(G(R))\}$$

and

$$\omega(G) = \omega(G(N)) + max\{\omega(G(A)), \omega(G(R))\}.$$

Proof Because $A \sim N$, from Proposition 1, it follows that

$$\alpha(G) = \alpha(G(A)) + max\{\alpha(G(N)), \alpha(G(R))\}.$$

Because $A \sim N \sim R$, it follows that

$$\omega(G) = \omega(G(N)) + max\{\omega(G(A)), \omega(G(R))\}.$$

5 Conclusions and Future Work

In this chapter we characterize weakly quasi-threshold graphs using the weakly decomposition, determine: density and stability number for weakly quasi-threshold graphs. Our future work concerns we give some applications of weakly quasi-threshold graphs including the medicine. Also we will explore the connection of weakly quasi-threshold graphs with the intelligent systems.

Acknowledgments This research was supported by the project entitled *Classes of graphs, complexity of problems and algorithms* in Bilateral Cooperation by Romanian Academy ("Vasile Alecsandri" University of Bacău is partner) and the National Academy of Sciences of Belarus and Belarusian Republican Foundation for Fundamental Research.

References

1. Corneil, D.G., Lerchs, H., Stewart, L.K.: Complement reducible graphs. Discrete Appl. Math. **3**, 163–174 (1981)
2. Bapat, R.B., Lal, A.K., Pati, S.: Laplacian spectrum of weakly quasi-threshold graphs. Graphs and Combinatorics **24**, 273–290 (2008)
3. Corneil, D.G., Perl, Y., Stewart, L.K.: A linear recognition algorithm for cographs. SIAM J. Comput. **14**, 926–934 (1985)
4. Brandstadt, A., Le, V.B., Spinrad, J.P.: Graph classes: a survey, SIAM Monographs on, Discrete Mathematics and Applications (1999).
5. Golumbic, M.C.: Algorithmic graph theory and perfect graphs. Second edition. Annals of Discrete Mathematics, vol. 57, Elsevier, Amsterdam (2004).
6. Berge, C.: Graphs. Nort-Holland, Amsterdam (1985)
7. Croitoru, C., Talmaciu, M.: A new graph search algorithm and some applications, presented at ROSYCS 2000, Alexandru Ioan Cuza University of Iași (2000).

8. Talmaciu, M.: Decomposition problems in the graph theory with applications in combinatorial optimization. PhD. Thesis, Alexandru Ioan Cuza University of Iasi, Romania (2002).
9. Talmaciu, M., Nechita, E.: Recognition algorithm for diamond-free graphs. Informatica **18**(3), 457–462 (2007)
10. Croitoru, C., Olaru, E., Talmaciu, M.: Confidentially connected graphs, The annals of the University "Dunarea de Jos" of Galati, Suppliment to Tome XVIII (XXII) 2000. In: Proceedings of the international conference "The risk in contemporany economy" (2000).
11. Nikolopoulos, S.D., Papadopoulos, C.: A simple linear-time recognition algorithm for weakly quasi-threshold graphs. Graphs and Combinatorics **27**, 567–584 (2011)

Large Graphs: Fast Cost Update and Query Algorithms. Application for Emergency Vehicles

Ion Cozac

Abstract This chapter presents a method that can be used to solve the shortest path problem in large graphs, together with arc cost updates. Our approach uses a contracted graph, which is obtained from the important nodes of the original graph. Every non-important vertex has one or more assigned important nodes as references. A reference node will help us to quickly find the arcs to be updated. The advantage of our method is that we can quickly update the contracted graph, so it can be safely used for future queries. An application of these algorithms can be used by emergency vehicles.

Keywords Shortest path problem · Contracted graph · Reference node · Emergency vehicle

1 Introduction

Determining a shortest path in large graphs is a problem that arises in many real world applications such as route planning for road networks or using train time-table information. A simple implementation of Dijkstra's algorithm [1] may need too much time to find the solution on large graphs such as the road network of a medium or big country, which may have millions of nodes and arcs. In order to reduce the amount of computations, many speed-up techniques have been developed during the last years. Every speed-up technique uses auxiliary data structures which are generated from the original graph in a pre-processing phase [2].

The pre-processing phase is executed once and for all, since it is supposed the graph topology is stable for long time. However, real road networks may change

I. Cozac (✉)
"Petru Maior" University of Tirgu-Mures, Târgu Mures, Romania
e-mail: cozac@upm.ro

B. Iantovics and R. Kountchev (eds.), *Advanced Intelligent Computational Technologies and Decision Support Systems*, Studies in Computational Intelligence 486, DOI: 10.1007/978-3-319-00467-9_19, © Springer International Publishing Switzerland 2014

sometimes, for example due to various unexpected events like traffic jams, natural disasters, construction works etc.

The *A* algorithm* [3] was one of the first speed-up technique which has been proposed to improve the performances of Dijkstra's algorithm by adding a potential to the priority of each node. The *bidirectional Dijkstra* [4] is a frequent speed-up technique used to solve the shortest path problem. Some speed-up techniques combine these two approaches.

The *ALT algorithm* [4] uses a small number of landmarks (usually 16 or 20), which are selected nodes using various criteria. The pre-processing phase computes distances between landmarks and all vertices of the graph. A particular search query is based on the *A* algorithm* and obtains the potential from pre-computed distances. This approach uses the following properties of distances: $d(v, L) - d(t, L) \leq d(v, t)$ and $d(L, t) - d(L, v) \leq d(v, t)$.

Landmarks may also be used for dynamic graphs [5]. In many cases, the Landmark-based approach may be used without rerunning the pre-processing phase if some arcs change their costs.

The *Arc-Flags* algorithm [6] partitions the graph into cells and attaches a label on each arc. A label contains a flag for each cell indicating whether a shortest path is starting from this arc to the corresponding cell. The *SHARC algorithm* [7] extends the Arc-Flags by using contraction: it iteratively removes non-important vertices and adds arcs between remaining nodes.

The *Highway hierarchies* approach [8] tries to exploit the hierarchy of a graph. A highway hierarchy is a set of graphs $G_0, \ldots, G_l$ where the number of levels $l + 1$ is given. The basic idea is to define a neighbourhood for each node which is defined by its closest neighbours. An arc (v, w) is of highway type if there is some shortest path $\langle s, \ldots, v, w, \ldots, t \rangle$ such that neither v is in the neighbourhood of t nor w is in the neighbourhood of s. After removing low degree nodes, the same procedure is applied recursively. On each level, the procedure creates shortcut arcs for each path containing by passable vertices.

A special case of the above approach is *Contraction Hierarchies* [9]. This approach is based on the concept of contraction, and the vertices are first ordered by "importance". A hierarchy is generated by iteratively contracting the least important node v, this means that a shortest path passing through v is replaced by a shortcut.

The *Reach-based routing* [10, 11] uses the following notion. The reach of a vertex v with respect to a path P from s to t, denoted by $r(v, P)$, is the minimum between the cost of subpath $\langle s, \ldots, v \rangle$ and the cost of subpath $\langle v, \ldots, t \rangle$. The reach of v is the maximum over all shortest paths through v of $r(v, P)$.

The *Transit node routing* approach [12, 13] is based on this heuristic: when we start from a source node and drives to somewhere "far away", we will leave the current location via one of only few "important" traffic junctions, which are called transit nodes. The pre-processing phase computes distances from each node to all neighbouring transit nodes and between all transit nodes. Non-local shortest path queries use a small number of table lookups. The selection of "important" nodes is usually based on highway hierarchies.

The *Highway node routing* approach [14] uses a multilevel overlay graph $(G_0,\ldots,G_l)$ such that $G_i = (V_i, A_i)$ where $A_i := \{(s,t) \in V_i \times V_i | \exists$ shortest path $P = \langle s, v_1, \ldots, v_k, t \rangle$ in G_{i-1} such that $\forall i : v_i \notin V_i\}$. The highway-node set is chosen by following this intuition: a node that lies on many shortest paths should belong to the node of a high level. The multi-level overlay graph is built in a bottom-up fashion.

The *Highway node routing* approach is also used to update arc costs. An arc cost update is solved separately for server scenario and mobile scenario, respectively. In the former, the server must update the auxiliary data structures such that any point-to-point query will be solved correctly. In the latter, the he expensive update step is skipped and they directly perform "prudent queries" that takes the changes into account.

Our chapter is structured as follows. Section 2 defines some basic notions and depicts the A* algorithm, which has been used for our implementation. Section 3 depicts the method we use to build two auxiliary data structures: the contracted graph and the list of references. Section 4 shows how to use the auxiliary data structures to find a shortest path between two distinct vertices. We also show how to increase or decrease an arc cost on both the original graph and on the contracted graph. Section 5 presents some experiments based on our approach. The last section is reserved for conclusions and discussions about future work, and includes also some remarks about how the medical system can use the algorithms described in this chapter.

Our approach starts with the built of the contracted graph using some simple techniques that allow us to reduce the number of vertices and arcs. For each non-important vertex that will not appear in the contracted graph, we identify one or more important nodes that can be used to quickly reach it. The references will help us to guide the search for shortest path if the start point and/or the target point are not important. The references will also help us to quickly detect the arcs of the contracted graph that have to be updated. The search algorithm uses the original graph and the contracted graph.

The contribution of this chapter is the usage of reference nodes to guide the search in the original graph, yielding very good response time for arc cost update operations.

2 Preliminaries

Graphs and paths. Let $G = (V, A)$ be a directed, weighted graph with $n = |V|$ nodes (vertices), $m = |A|$ arcs, and each arc (v, w) has a cost $c(v, w) > 0$. We assume the graph G is strongly connected, that is, there is a path from any vertex v to any other vertex w. The cost of a path $P = \langle v_1, \ldots, v_k \rangle$ is the sum of the cost of all its arcs:

$$c(P) = \sum_{i=2}^{k} c(v_{i-1}, v_i) \tag{1}$$

$P* = \langle s, \ldots, t \rangle$ is a shortest path from s to t if there is no other path P' such that $c(P*) > c(P')$. The distance $d(s, t)$ from s to t is the cost of a shortest path from s to t.

G is symmetric if every arc (v, w) has its pair (w, v) and $c(v, w) = c(w, v)$.

The $A*$ *algorithm* (see Fig. 1) is designed to determine a shortest path from s to t, that is, it is a point-to-point shortest path algorithm. For each vertex v, this algorithm computes a lower bound of the distance $d(v, t)$. Modern road networks include some supplementary information such as latitude and longitude of each vertex. If $d(v, t)$ is used to indicate the physical distance from v to t, we can use the euclidian distance between v and t as lower bound of $d(v, t)$. Let us denote this lower bound by $e(v, t)$—e may mean estimation or euclidian distance. If $d(v, t)$ is used to indicate the time needed to cover the path from v to t, we have to know the maximum possible speed to estimate a lower bound of $d(v, t)$.

In order to guarantee the correctness of the A* algorithm, the function e must be consistent, that is, for each arc (v, w) the following condition must be fulfilled:

$$e(v, t) \leq c(v, w) + e(w, t) \tag{2}$$

The euclidian distance always fulfils the above condition.

The function *Modify* modifies the assigned cost of the input vertex and organizes the priority queue Q using the information stored in vectors D and E. $D[w]$ is the cost of the current path from s to w via v, and $E[w]$ is a lower bound for a path from s to t, if this path uses the arc (v, w). The function *Extract* extracts the vertex that gives the best (minimum) estimation of the cost of a path from s to t. The vector P stores information that will be used to retrieve the path from s to t: $P[w] = v$ means that v is the predecessor of w in a path from s to t.

Fig. 1 The A* algorithm

```
begin
    for each v∈ V do D[v] := ∞;
    D[s] := 0;  E[s] := e(s,t);
    Q := {s};  Modify(Q,s,D,E);
    while (Q ≠ ∅) do begin
        v := Extract(Q,D,E);  if (v = t) then break;
        for each w ∈ Successors(v) do begin
            z := D[v] + c(v,w);
            if (D[w] > z) then begin
                if (w ∉ Q) then Q := Q∪{w};
                D[w] := z;  E[v] := z + e(w,t);
                P[w] := v;  Modify(Q,w,D,E);
            end
        end
    end
end (algorithm)
```

The priority queue Q must be implemented as binary heap of Fibonacci heap in order to achieve good response time, which gives $O(n \log n)$ time complexity order, assuming that the input graph is sparse—this assumption is true for road networks, which are planar. In fact, the response time depends heavily on the number of covered arcs, so this is a motivation to generate a contracted graph that "summarizes" the input one. The smaller graph will be used for rough searches, while the original graph will be used for refined searches.

3 Pre-processing Phase

First of all, we assume there are no self-loops (v, v) or parallel arcs (v, w); if it is the case, such arcs may be dropped in a pre-processing phase.

Our approach is based on the idea that, in a road network, we can identify a lot of nodes that have at most two neighbours. Now let us draw a road graph—which is clearly planar, and suppose the nodes that have at most two neighbours are "fading"—see Fig. 2. We obtain a contracted graph which has the same shape as the original one, but fewer nodes and arcs.

The nodes that define the new graphs are called *important nodes*. A node v is important if it has at least three neighbours. In other words, the vertex v has:

- either at least one incoming arc (x, v), and at least two outcoming arcs (v, y) and (v, z);
- or at least two incoming arcs (x, v) and (y, v), and at least one outcoming arc (v, z).

x, y and z are distinct vertices. We don't consider the other possible cases because it is assumed the input graph is strongly connected.

The first step of the pre-processing phase identifies the important nodes using the above criterion. One graph traversing is enough to determine, for every vertex v, if it has at most two or at least three neighbours. This step can be done in linear time. The important nodes will be used to build the contracted graph.

Every non-important vertex lies on a single line. A *line* between two important nodes s and t is a path $\langle s, v_1, \ldots, v_k, t \rangle$ where v_i are non-important vertices, that is,

Fig. 2 The original graph (**a**) is contracted and a smaller graph (**b**) is obtained

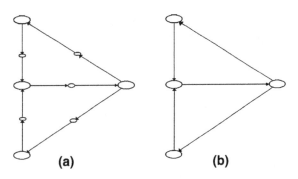

v_i has only two neighbours. There may be two or more disjoint paths from s to t; in this case we say the paths are *parallel*.

The second step of the pre-processing phase must build the arcs of the contracted graph and their costs. We also have to identify the reference nodes for all non-important vertices. Let us examine Fig. 3: a and b are important nodes, x and y are on the line between a and b. x can be reached only starting from a; y can be reached either starting from a or b. That is, an important node a is reference for x if there is a path from a to x that doesn't use any other important node.

To start this step of contraction, we initiate a breadth first search that starts from each important node and ends when another important node is reached. When the search reaches another important node, an arc of the contracted graph is created. If two or more parallel arcs from v to w (v and w are important nodes) are created, we keep only the shortest arc. The contracted graph should be scanned using the same procedures as for the original one. Sometimes a new contraction step needs to be made, for example if the current contracted graph contains parallel lines that link two important nodes.

When this type of contraction is no more possible, it is time to examine other possibilities that allow us to reduce the number of vertices and arcs. Let us examine Fig. 4. In the first case (a), an important node x has two predecessors and one successor. x will no more lie on the contracted graph and will have two assigned references: nodes a and b. All the non-important vertices that lie on the path from x to c will inherit the new references of x.

In the second case (b), an important node x has one predecessor and two successors. x will become non-important vertex and will have one assigned reference: node a. All the non-important vertices that lie on the paths $\langle x, \ldots, b \rangle$ and $\langle x, \ldots, c \rangle$ will inherit the reference of x.

We depicted above some contraction techniques for directed graphs, but we can adapt them to be used for symmetric graphs. The third case (c) depicts another contraction technique that can be used for symmetric graphs: solving the "delta" groups. A "delta" group is defined by five distinct vertices a, b, c, d and e with these properties:

- b has three neighbours: a, c and d;
- c has three neighbours: b, d and e;
- d has three neighbours: b, c and f.

In this case we can eliminate the inner vertices b, c and d together with their adjacent edges, to create six new arcs (three new edges) with the remaining important nodes a, e and f. This step must be executed only if the group $\{b, c, d\}$ follows the triangle inequality; otherwise, for example if $c(b, c) > c(b, d) + c(c, d)$,

Fig. 3 How to determine the reference nodes for non-important vertices

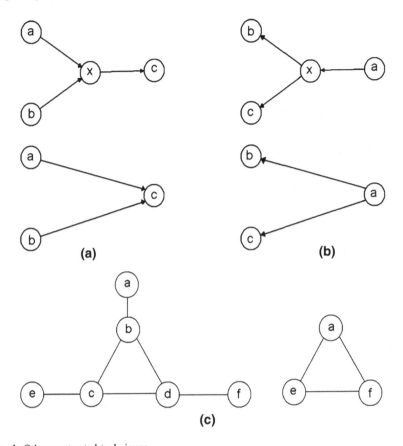

Fig. 4 Other contracted techniques

we simply drop the edge (b, c) and then rerun the contraction step to eliminate the vertices b and c together with their adjacent edges.

We don't investigate other cases for symmetric graphs because the number of edges will increase. Our future work will investigate some combinations of our approach with other methods, such as *Highway node hierarchy* or *SHARC*.

We may examine other possibilities to reduce the number of arcs: test each arc (v, w) to see if its dropping will give a new path P from v to w such that $c(P) < c(v, w)$, in this case we can drop this arc. The search for a new path should be stopped if a particular extracted vertex x has the assigned cost greater than $c(v, w)$, to avoid useless computations.

The final contracted graph has an important property. For each arc (v, w), its cost is equal to $d(v, w)$—the cost of a shortest path from v to w in the original graph.

How to represent the list of references for non-important vertices in a compact manner? Because one vertex may have only one assigned reference and another vertex may have two or more assigned references, we will use a method that is similar to the representation of a graph. The lookup table indicates, for each non-important vertex, where the list of its references starts in the reference list. The reference list enumerates, for each non-important vertex, its reference nodes.

4 Search for Shortest Path and Cost Arcs Update

4.1 Search for Shortest Path

The search process is divided into three steps. To obtain good response time, we must use the contracted graph, but what if s or t or both are not important? We have to perform the first search step using the original graph if s is not important, and one last search step if t is not important. The first search step will reach one or more important nodes that will be used as start points in the second search step. Using the contracted graph, the second search step will reach the reference nodes of the target vertex t. The last search step will finally reach the target vertex t.

We depicted above the worst case, if we have to run all these three search steps. But sometimes we may be lucky enough, so one or two search steps should be run to find a shortest path. We detail below the search algorithm.

First search step. If source node s is important, just insert it into the main priority queue with assigned cost zero. If s is not important, start an A* local search until either t or an important node is reached. If t is reached, the whole algorithm terminates. If an important node v is reached, insert it into the main priority queue Q with its computed cost, and continue the search algorithm without examining the successors of v.

Second search step. Before performing the search, we have to establish the stop criterion. The search will stop if either t or all the references of t are reached. Using a vector, we simply set the bits corresponding to these references. After the initialisation, the search mai start using only the contracted graph. If a reference node of t is reached, its successors are not examined, but the algorithm continues. The search step stops when the last reference of t is reached.

Third search step. Before performing this last step, we initialize the priority queue by inserting all the references assigned to t. This search step uses only the original graph.

Now we have to prove the correctness of this algorithm, that is, to prove that it finds a shortest path from s to t. Let us consider the most difficult case, with a shortest path containing at least two important nodes: $P = \langle s, \ldots, v, \ldots, w, \ldots, t \rangle$. v is the first important node of the path and has been reached from s; w is the last important node of the path and is a reference for t. First of all, it is clear that the subpaths $\langle s, \ldots, v \rangle, \langle v, \ldots, w \rangle$ and $\langle w, \ldots, t \rangle$ are optimum. The second path is optimum because every arc of the contracted graph corresponds to a shortest path in the original graph.

Suppose there is a path from s to t that is shorter than P and does not contain any important node. This hypothesis must be excluded because such a path should be detected in the first search step. So any shortest path should pass through an important node, which is reached in the first search step.

Suppose there is another path from s to t that is shorter than P, so that the subpath $\langle v, \ldots, t \rangle$ does not contain any other important node. This hypothesis must also be excluded because in the pre-processing phase the vertex t should be reached starting from v.

So the search algorithm depicted above determines a shortest path from s to t.

4.2 Arc Cost Update

If we have to update the cost of an arc (v, w), we use this simple strategy. Using the original graph, start a local A* search from each reference of v and compute the cost to each reached vertex. The search stops when another important node is reached. Of course, when the search encounters the arc (v, w), its cost will be updated in the original graph. When another important node is reached, the cost of the corresponding arc is updated if necessary. Our approach allows us to quickly update the arc cost of the contracted graph, so we don't need to perform "prudent queries" using a bigger graph.

5 Experiments

We implemented our algorithms in ANSI C and compiled them using GNU C Compiler version 4.1.0 on an AMD Athlon processor at 1,4 GHz with 2 GB of RAM running Linux Red-Hat 4.1.0-3.

We deal with a fictitious directed graph, which simulates a detailed road map used by mobile devices. The map contains information about streets, buildings and other interest objectives that are on the street.

The original graph has 7.960.800 vertices and 8.120.000 arcs, the contracted graph has 159.200 important vertices and 318.400 arcs. The reference list has 7.801.600 entries because we used only one reference for each non-important vertex. The contracted graph and the reference list took about 7 s to be generated. The original graph needs about 92.35 MB of memory to be stored, and all data structures need about 185,25 MB.

We consider two query data categories. The first category includes paths between extreme vertices at maximum distance, and the second category includes paths between random vertices at medium distance, related to the maximum distance found on the first category. We compare the performances of three algorithms: plain Dijkstra using only the original graph returning the detailed path, A* algorithm using contracted graph and/or the original graph and returning the resumed path, A*

Table 1 Performances of the evaluated algorithms

	Plain dijkstra	A* outputs returned path	A* outputs detailed path
Elementary operations	151,328,000	3,159,000	3,286,000
	94,462,000	867,000	936,000
Covered arcs	15,925,000	347,000	410,000
	9,819,000	96,000	131,000
Path length in number of arcs	20,350	840	20,350
	11,380	450	11,380
Response time in seconds	19.60	0.25	0.29
	12.80	0.08	0.11

For each cell, the above number corresponds to the first category, and the below number corresponds to the second category

algorithm returning the detailed path. The A* algorithm may return either the resumed path, that is, the path containing only important nodes (plus, if it is the case, the paths found by the first and the third steps), or the detailed path. In the second case, local A* searches are needed for each arc that links two important nodes, in order to find the detailed path containing all needed vertices.

For each search algorithm we determine the number of elementary operations that are executed by *Modify* and *Extract* functions (see Sect. 2), the number of covered arcs and the running time. An elementary operation is a comparison between two elements of the priority queue. We take into account this parameter because it is more relevant than the running time; in fact, the complexity of any algorithm depends on the number of its elementary operations needed to obtain the output result (Table 1).

The A* algorithm using the contracted graph is about 70–120 times faster than plain Dijkstra if we output the detailed path, and about 80–160 times faster if we output the resumed path. Our approach is not so powerful as the other methods mentioned in the introduction, but there is room for improvements. Practical applications may output the resumed graph and, if necessary, few A* local searches may be performed to output local detailed paths.

Another experiment evaluates the performance of the update algorithm. 8,000,000 independent arc cost updates have been performed in about 150 s, using 860,000,000 elementary operations and 840,000,000 covered arcs. Relating this information per arc cost update yields: 11 elementary operations, 11 covered arcs and 0.00001875 s. Our approach is the most efficient over all methods from update point of view.

6 Conclusions and Future Work

This chapter presented a speed-up technique that can be used to quickly find shortest path in large road networks. The algorithms can be easily implemented for mobile devices, including solutions for arc cost updates.

These algorithms can be used in the medical system, for example to optimize the traffic of the emergency vehicles. A customized application may be designed to watch over this traffic, if the vehicles are endowed with GPS devices. If some portions of the road network is changed, for example due to various unexpected events like traffic jams, natural disasters, construction works etc., the driver of the emergency vehicle may easily update the graph, so he will quickly find a new optimum route with respect to the new information.

There is room to improve the performance if our method is combined with other approaches, such as *Highway node hierarchies, Contraction Hierarchies* or *SHARC*. Extending our approach will give much better response time for search path queries, while the pre-processing phase and arc cost updates may slow down. This should not be a problem: the pre-processing phase is executed once and for all, and search path queries are much more frequent than arc cost updates, so the extension is justified. However, we should find equilibrium between the number of references allowed for each non-important vertex and the parameters of other methods, to guarantee both acceptable memory consumption and response time for queries.

Our approach is valid for graphs with stable topology and all the information is done at the beginning. However, new approaches are needed for very dynamic graphs, where new arcs may be inserted and other may be dropped frequently. More than that, we may have no information about the whole graph, but only for the neighbours of few vertices. This approach is very useful for MANETs— Mobile Ad-hoc NETworks [15].

Acknowledgments This work was supported by the Bilateral Cooperation Research Project between Bulgaria-Romania (2010–2012) entitled *"Electronic Health Records for the Next Generation Medical Decision Support in Romanian and Bulgarian National Healthcare Systems"*, *NextGenElectroMedSupport*.

References

1. Dijkstra, E.W.: A note on two problems ion connection with graphs. Numer. Math. **1**, 269–271 (1959)
2. Bauer, R., Delling, D., Wagner, D.: Experimental study of speed-up techniques for timetable information systems. J. Netw. **57**, 38–52 (2011)
3. Hart, P.E., Nilsson, N., Raphael, B.: A formal basis for the heuristic determination of minimum cost paths. IEEE Trans. Syst. Sci. Cybern. **4**, 100–107 (1968)
4. Goldberg, A., Harelson, C.: Computing the shortest path: A* search meets graph theory. In: Proceedings of 16th Annual ACM-SIAM Symposium Discrete Algorithms (SODA'05), pp. 156–165 (2005)
5. Delling, D., Wagner, D.: Landmark-based routing in dynamic graphs. In: Proceedings of 6th Workshop of Experimental Algorithms, Lecture Notes in Computer Science, vol. 4525, pp. 52–65. Springer, June 2007
6. Möhring, R.H., Schilling, H., Schiltz, B., Wagner, D., Willhalm, T.: Partitioning graphs to speed up Dijkstra's algorithm. In: Proceedings of 4th Workshop on Experimental Algorithms (WEA'05), Lecture Notes in Computer Science, vol. 3503, pp. 189–202. Springer, 2005

7. Bauer, R., Delling, D.: SHARC: fast and robust unidirectional routing. In: Proceedings of 10th Workshop of Algorithms and Engineering Experiments (ALENEX'08), SIAM, pp. 13–26 April 2008
8. Sanders, P., Schultes, D.: Engineering highway hierarchies. In: Proceedings of 14th Annual European Symposium Algorithms (ESA'06), Lecture Notes in Computer Science, vol. 4168, pp. 804–816. Springer, 2006
9. Geisberger, R., Sanders, P., Schultes, D., Delling, D.: Contraction hierarchies: faster and simpler hierarchical routing in road networks. In: Proceedings of 7th Workshop Experimental Algorithms, pp. 319–333 (2008)
10. Goldberg, A., Kaplan, H., Werneck, R.F.: Reach for A*: efficient point-to-point shortest path algorithms. In: Proceedings of the 8th Workshop on Algorithms Engineering and Experiments (ALENEX'06), SIAM, pp. 129–143 (2006)
11. Gutman, R.J.: Reach-based routing: a new approach to shortest path algorithm optimized for road networks. In: Proceedings of 6th Workshop Algorithm Engineering and Experiments (ALENEX'04), SIAM, pp. 100–111 (2004)
12. Bast, H., Funke, S., Matijevic, S., Sanders, P., Schultes, D.: Intransit to constant time shortest path queries in road networks. In: Proceedings of the 9th Workshop on Algorithm Engineering and Experiments and the 4th Workshop on Analytic Algorithmics and Combinatorics, pp. 46–59 (2007)
13. Bast, H., Funke, S., Sanders, P., Schultes, D.: Fast routing in road networks with transit nodes. Science **316**, 566 (2007)
14. Schultes, D., Sanders, P.: Dynamic highway-node routing. In: Proceedings of 6th Workshop on Experimental Algorithms (WEA'07), Lecture Notes in Computer Science, vol. 4525, pp. 66–79. Springer, June 2007
15. Chlamtac, I., Conti, M., Liu, J.: Mobile ad-hoc networking: imperatives and challenges, pp. 13–64. Ad-Hoc Networks I, Elsevier (2003)

Scan Converting OCT Images Using Fourier Analysis

Amr Elbasiony and Haim Levkowitz

Abstract Scan conversion is the process by which a polar image is transformed into Cartesian coordinates. Several image modalities such as radar and catheter based imaging modalities acquire data in a polar image format. While suitable for the acquisition process, this format poses a challenge both in displaying as well as on the analyses of the image. Intravascular optical coherence tomography (IVOCT) is a catheter-based imaging modality utilizing a polar image format. Current interpolation techniques used to fill in the uneven spacing between the lines of IVOCT images create visual artifacts and render the resulting image unsuitable for reliable analysis. We present a novel technique that estimates the unsampled pixel values between the lines of IVOCT images using the Fourier analysis of the acquired data. This technique should minimize the visual artifacts as well as provide images more reliable for further analysis.

Keywords Digital scan conversion · Polar image analysis · IVOCT decision support systems · IVOCT speckle analysis

1 Introduction

Images in polar format are usually acquired by a rotating imaging device such as a radar or a catheter. A 2D polar image is acquired line by line. Each line is acquired at an equal angular increment as a sequence of data points from the imaging device to the maximum imaging distance. Additional lines are acquired at different angular positions as the imaging device rotates around one of its axis in the same

A. Elbasiony (✉) · H. Levkowitz
Univeristy of Massachusetts Lowell, Lowell, USA
e-mail: aelbasio@cs.uml.edu

H. Levkowitz
e-mail: haim@cs.uml.edu

B. Iantovics and R. Kountchev (eds.), *Advanced Intelligent Computational Technologies and Decision Support Systems*, Studies in Computational Intelligence 486, DOI: 10.1007/978-3-319-00467-9_20, © Springer International Publishing Switzerland 2014

plane. The resulting image has a varying spatial resolution in the angular direction with respect to the distance from the imaging device. The further from the device the lower the spatial resolution, as shown in Fig. 1.

A transformation from polar format into Cartesian coordinates is often desirable for proper viewing as well as analysis of the images. In addition, most display systems, such as computer screens are designed and driven by graphics hardware based on Cartesian coordinates and hence expect the position of pixels to be specified in the Cartesian (x, y) coordinates not in the polar (r, θ) coordinates.

The direct transformation from polar to Cartesian coordinates produces images with unacceptable quality due to the radially increasing gaps between the lines, as shown in Fig. 2.

As an example, one of the currently available commercial OCT systems has 960 data points per scan line and 504 scan lines per frame. This gives a total of $504 \times 960 = 483,840$ data points. The region occupied by a circle of radius 960 is $\pi \times (960)^2 = 2,893,824$ pixels, which account for more than five times the number of acquired data points. This large difference is typically occupied by unsampled data points that need to be estimated. Even though a large percentage of the final image is made of estimated data, it is important to note that the estimation of data is only practical when applied to scales much smaller than the scales of the physiological features presneted in the image. If the original image resolution is too low to the point that unsampled data are estimated between different types of physiological features at scales comparable to the size of the phsiological features themseleves, then no estimation technique would work without creating false structures that render the final image unusable. The new technique presented in

Fig. 1 Illustration of polar image acquisition

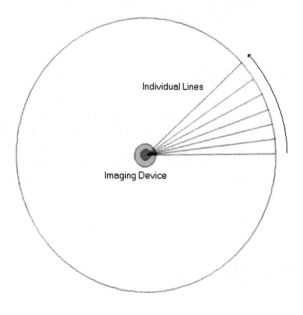

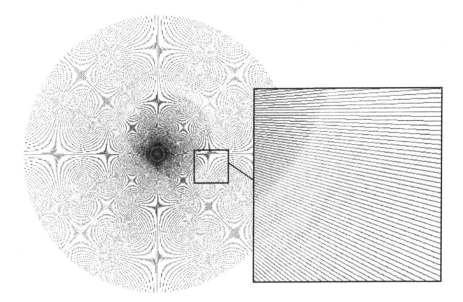

Fig. 2 Simple polar-to-raster transformation

this paper assumes the estimation of unsampled data is limited to scales much smaller than (typically at the speckle scale) the physiological features in the image.

Digital scan conversion is the process by which a polar image is transformed into Cartesian coordinates. Most algorithms developed for digital scan conversion rely on some form of interpolation in order to estimate the unsampled pixel values between the lines. Some of the earliest attempts utilized the nearest-neighbor interpolation technique. While fast and suitable for the computational capacity available at the time, it has been shown by Stark et al. [1] to produce both high- and low-frequency artifacts. As the computational capacity of available computers increased, more computationally demanding techniques, such as bilinear, bicubic, and high-resolution cubic spline interpolation were suggested. The work of Parker et al. [2] provided a comparison between different types of interpolation functions and concluded that the high-resolution cubic spline function provided better image quality at the expense of increased computational time. Nonetheless, bilinear and cubic interpolation algorithms suffer from the smoothing artifact that can blur edges and decrease image details. This problem has been addressed in the literature by different adaptive interpolation techniques [3–5]. Ma et al. [6] proposed the Kriging interpolation algorithm to distinguish detail regions and smooth regions with asymmetry operator. Recently, Ahn et al. [7] suggested a scan conversion method in the frequency domain using Fourier transform. While one would expect their method to be comparable to the nearest-neighbor interpolation, they argued that by using Kaiser filtering it yielded comparable results to the bilinear interpolation at the expense of more computational time.

Fig. 3 Illustration of the intrinsic dependence on distance in polar images

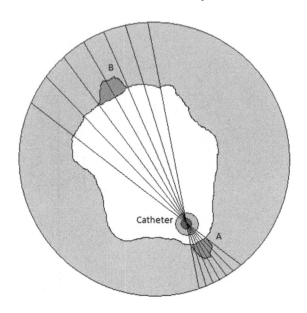

Not only the polar image format presents a challenge in displaying the image, it also provides a major obstacle when attempting to apply region-based image analysis. While the consistency of quantitative measures plays a pivotal role in any quantitatively based image analysis, region-based measures extracted from polar images produce inconsistent results that are highly dependent on the radial location they have been extracted from. For instance, consider Fig. 3, if region A and region B both represent the same imaged object, the spatial resolution along the radial direction on each scan line is the same in both regions but the spatial resolution along the angular direction between scan lines is considerably different. This intrinsic inconsistency renderes region-based techniques essentially inapplicable to polar images.

2 Scan Conversion Using Fourier Analysis

All current interpolation schemes rely on the implicit (or explicit) assumption that a given intensity value changes continuously and incrementally between the lines. This assumption is generally far from reality in the presence of speckles as in the case of OCT images. In this type of imaging modalities, intensity values are expected to change abruptly due to interference between coherent waves used in the imaging process, as explained by Goodman [8]. The simple interpolation of these abrupt changes all the way between the lines produces smear- or elongation artifacts, as shown in Fig. 4. The speckle patterns in the figure become more stretched as the gap between the lines increases.

Fig. 4 Scan conversion
using bilinear interpolation

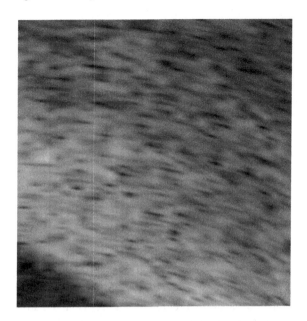

To produce a more consistent image and avoid the elongation artifact, we rely on a different assumption. We assume that the behavior of intensity changes in the radial dimension can be extended symmetrically into the angular dimension. In other words, abrupt changes in the radial dimension should appear as abrupt changes in the angular dimension as well while, slow changes in the radial dimension should also appear as slow changes in the angular dimension.

To implement a scan conversion method based on our new assumption, we resort to a mathematical utility, Fourier analysis, to analyze the radial dimension intensity distribution in a given local neighborhood and estimate the missing pixel values by properly extending the analysis result to the angular dimension. Figure 6 provides an illustration of the idea. Small circles in the figure represent high frequency waves while large circles represent low frequency waves. The intensity value at any point between the two lines is given by the inverse Fourier transform of the superposition of these individual waves from both lines. The Fourier transform period is assumed to be variable and centered around the line through the point of interest. The higher the frequency the smaller the period width and the lower the frequency the larger the period width. The maximum period width equals to the width between the lines at the point of interest, as shown in Fig. 5.

The high frequencies corresponding to abrupt transitions in the intensity values will have small width and hence will die quickly close to the line while the low frequencies will span more distance toward the other line. This results in a normalized, naturally looking speckle pattern without the artifacts introduced by the interpolation methods.

The unsampled data at a sample position x between any two successive scan lines L_1 and L_2 is computed as a line segment connecting sample position x in L_1

Fig. 5 Selecting the discrete
Fourier transform period

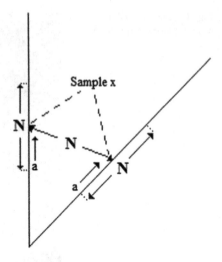

with sample position x in L_2 as shown in Fig. 5. If the line segment has width
N then we analyze a region of width N centered around sample position x on each
of the two scan lines. Each of the two scan lines contributes half of the values in
the line segment.

The one dimension discrete Fourier transform for a given frequency f is given
by:

$$U_f = \sum_{n=0}^{N-1} I_n \exp \frac{-2\pi i f n}{N} \tag{1}$$

where N is the width of the Fourier window (the distance between the lines at a
given sample position) and I_n is the intensity value at position n within the

Fig. 6 Illustration of change
symmetry in 2D

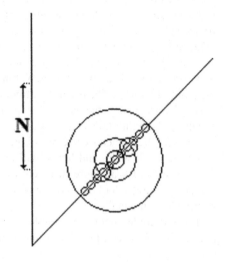

window. The superposition of the Fourier transform for a given frequency f, is given by:

$$U_f = \sum_{n=0}^{N-1} I_n \exp \frac{-2\pi \; if \; n}{N} + \sum_{n=0}^{N-1} I_n \exp \frac{-2\pi \; if \; (N-n)}{N} \quad (2)$$

Notice that for the second line we calculate the transform using N-n as opposed to n. This ensures that the result of the inverse transform will localize these changes around their respective lines. Finally, the complete formulation when the Fourier window is centered around the sample position is given by:

$$U_f = \sum_{n=-k}^{+k} I_a \exp \frac{-2\pi \; if \; |n|}{N} + \sum_{n=-k}^{+k} I_a \exp \frac{-2\pi \; if \; (N-|n|)}{N} \quad (3)$$

where, $k = \frac{N}{2}$, and $a = \frac{N}{2} + n$.

The implementation of the method can be better explained via the Euler's formula:

$$\exp ix = \cos x + i \sin x \quad (4)$$

The following pseudo-code illustrates the implementation of Eq. 3

```
1:  for f = 0 to N do
2:      Re[f] ← 0
3:      Im[f] ← 0
4:      for n = -N/2 to N/2 do
5:          a ← N/2 + n
6:          t₁ ← (-2π|n|)/N f
7:          t₂ ← (-2π(N-|n|))/N f
8:          Re[f] ← Re[f] + I₁(a) cos(t₁) + I₂(a) cos(t₂)
9:          Im[f] ← Im[f] + I₁(a) sin(t₁) + I₂(a) sin(t₂)
10:         n ← n + 1
11:     end for
12:     f ← f + 1
13: end for
```

where Re is the real part, Im is the imaginary part, $I_1(a)$ is the intensity value taken from one scan line at location a within the transform window N centered around a given sample position x, and $I_2(a)$ is the corresponding intensity value taken from the next scan line at the same sample position. The result of the transform is a one dimension frequency domain of a line segment of width N between the two scan lines.

When computing the inverse Fourier transform we enforce the model described above by defining a span window s for any given frequency f. The maximum frequency for a given window of size N is at the Nyquist frequency $N/2$. According to our model, the window at the Nyquist frequency should have width 1 allowing abrupt per-pixel changes to appear around their corresponding lines and masked everywhere else. As the frequency decreases its span window increases allowing corresponding intensity variations to span more distance from the lines as illustrated in the following pseudo-code:

```
 1: NyquistFrequency ← N/2
 2: for n = 0 to N do
 3:    Iₙ ← 0
 4:    for f = 0 to N do
 5:       t ← (2πn/N) f
 6:       s ← |NyquistFrequency − f| + 1
 7:       if n < s or n > (N − s) then
 8:          Iₙ ← Iₙ + (Re[f] cos (t) − Im[f] sin (t))
 9:       end if
10:       f ← f + 1
11:    end for
12:    Iₙ ← (Iₙ/2N) {Normalize}
13:    n ← n + 1
14: end for
```

3 Methods and Material

A hollow cylindrical uniform phantom material was imaged by the C7XR OCT system.[1] To image a segment of the phantom material, a series of cross-sectional frames is acquired by moving the imaging catheter longitudinally within the phantom. This is typically performed automatically by the imaging system to ensure constant speed during the acquisition process. An imaging catheter was inserted inside the hollow phantom and its imaging tip was positioned right outside the phantom. The acquisition process was then started and the catheter was automatically pulled inside the phantom. Any acquired frames outside the phantom material were excluded from our analysis. Only frames acquired inside the phantom material were considered usable. The imaging process was repeated five times with five different catheters to account for setup and manufacturing variabilities. The acquired polar data was scan converted using the method described above then smoothed using 5×5 Gaussian filter to reduce the effect of noise. The first and last frame of usable data were identified then thirty frames within the range of usable data were randomly selected for each of the five acquired phantom cross-sectional segments for a total of 150 frames. Several regions were manually selected on each frame at different distances from the catheter for a total of 1,325 regions. The normalized variance (variance divided by the mean intensity value) and entropy were calculated for each selected region. The same calculations were

---

[1] LightLab Imaging, Westford, MA.

also performed on the same images scan converted using bilinear interpolation and smoothed using the same Gaussian filter for comparison.

$$\sigma^2 = \frac{1}{N}\sum_{i=1}^{N}(I_i - \mu)^2 \tag{5}$$

where σ^2 is the variance. N is the number of intensity values in the selected region. I_i is the i_{th} intensity value and μ is the mean intensity given by:

$$\mu = \frac{1}{N}\sum_{i=1}^{N}I_i \tag{6}$$

$$E = -\sum_{i=1}^{N}p\,(I_i)\log p\,(I_i) \tag{7}$$

where E is the entropy and p is the probability of intensity value I_i.

4 Results and Discussion

Figure 7 presents the scan conversion results of a synthetic radial random blob image. Image (b) shows the scan converted image using bilinear interpolation where, the individual blobs on the radial dimension have been smeared into elongated artifacts on the angular dimension. By contrast, our new technique shown on image (a) preserved the individual blobs on the radial dimension and extended them into the angular dimension.

Additional comparison examples from real coronary artery images are shown in Fig. 8. On the left the speckle pattern in the scan converted images using our method no longer appear smeared or elongated compared with their corresponding counter parts scan converted using bilinear interpolation on the right.

Quantitatively, the use of the new scan conversion method has produced more normalized speckle patterns that significantly reduced the dependence on the distance compared to bilinear interpolation. As shown in Fig. 9, our method shows a more consistent results over different distances, in contrast, the bilinear interpolation clearly shows inconsistency as the distance from the catheter increases. The correlation coefficient between the normalized variance and the distance has decreased from 0.38 in the case of bilinear interpolation to -0.12 for our method.

The information content is assessed by measuring the entropy. Entropy has long been used to measure the amount of randomness of a given distribution. The larger the entropy, the less the statistical dependence and hence the less the order and correlation in a given region. Figure 10 shows consistent reduction in the entropy in images scan converted using our method compared to bilinear interpolation as

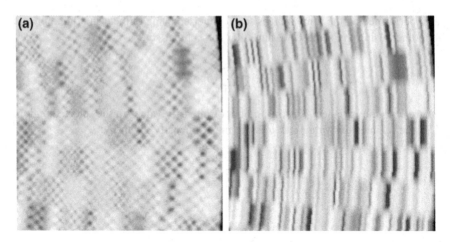

Fig. 7 Scan converted random blob image: **a** using our method and **b** using bilinear interpolation

we move further from the catheter. This is generally considered as an indicator to potential information gain. While the assessment of the significance of this potential information gain would need further analysis, our preliminary analysis provide promising results to motivate the pursue of further investigation.

In terms of computational complexity, our method has considerably higher computational complexity compared to bilinear interpolation. We performed all of our analysis using discrete Fourier transform. The use of FFT can speed up the process only when the transform window exceeds 5 pixels. This can easily be explained by comparing the computational complexity of DFT to that of FFT. The computational complexity of DFT is n^2, when $n = 5$ the complexity is 25. For FFT the computational complexity is $n \log_2 n$, when $n = 8$ (we need to pad to the nearest power of 2) the complexity is 24. For 504 lines per frame the width between the lines reaches 5 pixels at approximately sample position 401, which is about half way through the total 960 samples per line. Whether the DFT or the FFT is used, the technique is more computationally intensive compared to bilinear interpolation.

Finally, it is important to understand that the proposed method is not intended to recover the physically unsampled data, this is a profound claim that we do not attempt to make here. It merely analyzes real acquired data and create images with a more normalized speckle pattern using these analysis. We did not attempt to compare the resulting images with higher resolution images because such analysis would be irrelevant to this research. Nonetheless, for the interested reader, we offer the following suggestion. Acquiring two sets of data at two different points in time opens the door to many variables that would make it extremely difficult to carry out such analysis. Instead, we suggest acquiring the data at the highest resolution, then scan convert a subset of the lines using our method and compare the result to the complete set. The first obvious question would then be; to what higher resolution (if any) this scan converting technique would be comparable to? It can not possibly be comparable to all arbitrary higher resolutions.

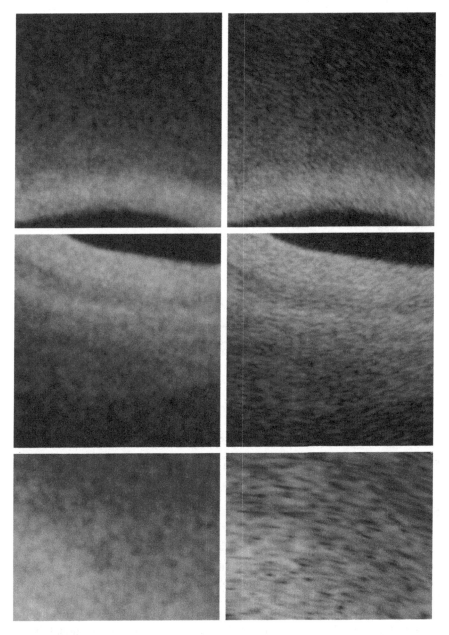

Fig. 8 Scan converted IVOCT arterial wall images. *Left* using our method and *Right* using bilinear interpolation

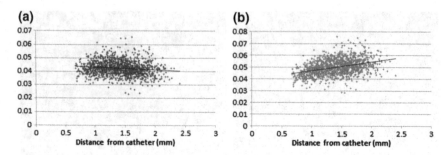

Fig. 9 Scatter plot of the normalized variance for **a** images scan converted using our method and **b** images scan converted using bilinear interpolation

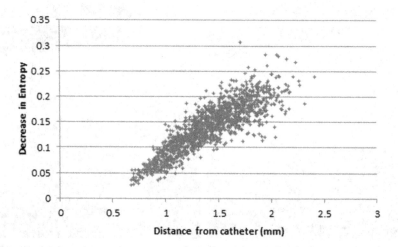

Fig. 10 Scatter plot of the potential information gain

5 Conclusion

In this paper a new scan conversion technique has been proposed in the scan conversion of OCT images, utilizing Fourier analysis. The new technique has been compared to the commonly used interpolation technique. Results show that the new technique enhanced the quality of the resulting images by significantly reducing the elongation and smear artifacts in the resulting speckle pattern compared to the interpolation technique at the expense of more computational time.

Furthermore, the resulting normalized speckle patterns produce more consistent quantitative results with less dependency on the distance from the catheter. In addition, the entropy analysis suggests a potential information gain, nonetheless, its significance has yet to be assessed.

References

1. Stark, H., Woods, J., Paul, I., Hingorani, R.: Direct fourier reconstruction in computer tomography. Acoustics, Speech and Signal Process., IEEE Trans. on **29**(2), 237–245 (1981)
2. Parker, J.A., Kenyon, R.V., Troxel, D.E.: Comparison of interpolating methods for image resampling. Medical Imaging, IEEE Trans. on **2**(1), 31–39 (1983). 10.1109/TMI.1983.4307610.
3. Hadhoud, M., Dessouky, M., Abd El-Samie, F., El-Khamy, S.: Adaptive image interpolation based on local activity levels. pp. C4–1-8 (2003). 10.1109/NRSC.2003.1217337.
4. Hwang, J.W., Lee, H.S.: Adaptive image interpolation based on local gradient features. Signal Process. Lett., IEEE **11**(3), 359–362 (2004). 10.1109/LSP.2003.821718.
5. Ma, L., Ma, J., Shen, Y.: Local activity levels guided adaptive scan conversion algorithm. pp. 6718–6720 (2005). 10.1109/IEMBS.2005.1616045.
6. Ma, L., Ma, J., Shen, Y.: Kriging interpolation based ultrasound scan conversion algorithm. pp. 489–493 (2006). 10.1109/ICIA.2006.305782.
7. Ahn, D.K., Jeong, M.K., Kwon, S.J., Bae, M.H.: A digital scan conversion algorithm using fourier transform. In: Ultrasonics Symposium (IUS), 2009 IEEE International, pp. 1322–1325 (2009). 10.1109/ULTSYM.2009.5441690.
8. Goodman, J.W.: Speckle Phenomena in Optics Theory and Application. Roberts & Company, Englewood (2007)

Printed in the United States
By Bookmasters